INTRODUCTION
TO
STOCHASTIC
PROCESSES

Chapman & Hall Probability Series

Series Editors
Jean Bertoin, Université Paris VI
David S. Griffeath, University of Wisconsin
Marjorie G. Hahn, Tufts University
J. Michael Steele, University of Pennsylvania

Hoffman-Jørgensen, J., *Probability with a view toward Statistics, Volumes I and II*

Stromberg, K., *Probabitlity for Analysts*

Lawler, Gregory F., *Introduction to Stochastic Processes*

INTRODUCTION TO STOCHASTIC PROCESSES

Gregory F. Lawler
Duke University

CHAPMAN & HALL

London · Weinheim · New York · Tokyo · Melbourne · Madras

Published by Chapman & Hall, 2-6 Boundary Row, London SE1 8HN, UK

Chapman & Hall, 2-6 Boundary Row, London SE1 8HN, UK

Chapman & Hall GmbH, Pappelallee 3, 69469 Weinheim, Germany

Chapman & Hall USA, 115 Fifth Avenue, New York, NY 10003, USA

Chapman & Hall Japan, ITP-Japan, Kyowa Building, 3F, 2-2-1 Hirakawacho, Chiyoda-ku, Tokyo 102, Japan

Chapman & Hall Australia, 102 Dodds Street, South Melbourne, Victoria 3205, Australia

Chapman & Hall India, R. Seshadri, 32 Second Main Road, CIT East, Madras 600 035, India

First edition 1995
Reprinted 1996

© 1995 Chapman & Hall

Printed in Great Britain by St Edmundsbury Press Ltd

ISBN 0 412 99511 5

A Catalogue record for this book is available from the British Library

Library of Congress Cataloging-in-Publication Data 94 36493 CIP

♾ Printed on permanent acid-free text paper, manufactured in accordance with ANSI/NISO Z39.48-1992 and ANSI/NISO Z39.48-1984 (Permanence of Paper)

Contents

Preface

This book is an outgrowth of lectures in Mathematics 240, "Applied Stochastic Processes," which I have taught a number of times at Duke University. The majority of the students in the course are graduate students from departments other than mathematics, including computer science, economics, business, biological sciences, psychology, physics, statistics, and engineering. There have also been graduate students from the mathematics department as well as some advanced undergraduates. The mathematical background of the students varies greatly, and the particular areas of stochastic processes that are relevant for their research also vary greatly.

The prerequisites for using this book are a good calculus-based undergraduate course in probability and a course in linear algebra including eigenvalues and eigenvectors. I also assume that the reader is reasonably computer literate. The exercises assume that the reader can write simple programs and has access to some software for linear algebra computations. In all of my classes, students have had sufficient computer experience to accomplish this. Most of the students have also had some exposure to differential equations and I use such ideas freely, although I have a short section on linear differential equations in the preliminary chapter.

I have tried to discuss key mathematical ideas in this book, but I have not made an attempt to put in all the mathematical details. Measure theory is not a prerequisite but I have tried to present topics is a way such that readers who have some knowledge of measure theory can fill in details. Although this is a book intended primarily for people with applications in mind, there are few real applications discussed. True applications require a good understanding of the field being studied and it is not a goal of this book to discuss the many different fields in which stochastic processes are used. I have instead chosen to stick with the very basic examples and let the experts in other fields decide when certain mathematical assumptions are appropriate for their application.

Chapter 1 covers the standard material on finite Markov chains. I have not given proofs of the convergence to equilibrium but rather have emphasized the relationship between the convergence to equilibrium and the size of the eigenvalues of the stochastic matrix. Chapter 2 deals with infinite state space. The notions of transience, null recurrence, and positive recurrence are introduced, using as the main example, a random walk on the nonnegative integers with reflecting boundary. The chapter ends with a discussion of branching processes.

Continuous-time Markov chains are discussed in Chapter 3. The discussion centers on three main types: Poisson process, finite state space, and birth-and-death processes. For these processes I have used the forward differential equations to describe the evolution of the probabilities. This is easier and more natural than the backward equations. Unfortunately, the forward equations are not a legitimate means to analyze all continuous-time Markov chains and this fact is discussed briefly in the last section. One of the main examples of a birth-and-death process is a Markovian queue.

I have included Chapter 4 on optimal stopping of Markov chains as one example in the large area of decision theory. Optimal stopping has a nice combination of theoretical mathematics leading to an algorithm to solve a problem. The basic ideas are also similar to ideas presented in Chapter 5.

The idea of a martingale is fundamental in much of stochastic processes, and the goal of Chapter 5 is to give a solid introduction to these ideas. The modern definition of conditional expectation is first discussed and the idea of "measurable with respect to $\mathcal{F}_n$, the information available at time n" is used freely without worrying about giving it a rigorous meaning in terms of σ-algebras. The major theorems of the area, optional sampling and the martingale convergence theorem, are discussed as well as their proofs. Proofs are important here since part of the theory is to understand why the theorems do not always hold. I have included a discussion of uniform integrability.

The basic ideas of renewal theory are discussed in Chapter 6. For non-lattice random variables the renewal equation is used as the main tool of analysis while for lattice random variables a Markov chain approach is used. As an application, queues with general service times are analyzed.

Chapter 7 discusses a couple of current topics in the realm of reversible Markov chains. First a more mathematical discussion about the rate of convergence to equilibrium is given, followed by a short introduction to the idea of Markov chain algorithms which are becoming very important in some areas of physics, computer science, and statistics. The final section on recurrence is a nice use of "variational" ideas to prove a result that is hard to prove directly.

Chapter 8 gives a very quick introduction to a large number of ideas in Brownian motion. It is impossible to make any attempt to put in all the mathematical details. I have discussed multidimensional as well as one-dimensional Brownian motion and have tried to show why Brownian motion and the heat equation are basically the same subject. I have also tried to discuss a little of the fractal nature of some of the sets produced by Brownian motion. In Chapter 9, a very short introduction to the idea of stochastic integration is given. This also is a very informal discussion but is intended to allow the students to at least have some ideas of what a stochastic integral is.

This book has a little more than can be covered in a one semester course. In my view the basic course consists of Chapters 1, 2, 3, 5, and 8. Which of the remaining chapters I cover depends on the particular students in the class that semester. The basic chapters should probably be done in the order listed, but the other chapters can be done at any time. Chapters 4 and 7 use the previous material on Markov chains; Chapter 6 uses Markov chains and martingales in the last section; and Chapter 9 uses the definition of Brownian motion as well as martingales.

I would like to thank the students in Math 240 in Spring 1992 and Spring 1994 for their comments and corrections on early versions of these notes. I also thank Rick Clelland, who was my assistant when I was preparing the first version in 1992, and the reviewers, Michael Phelan and Daniel C. Wiener, for their suggestions. During the writing of this book, I was partially supported by the National Science Foundation.

INTRODUCTION
TO
STOCHASTIC
PROCESSES

Preliminaries

0.1 Introduction

A *stochastic process* is a random process evolving with time. More precisely, a stochastic process is a collection of random variables X_t indexed by time. In this book, time will always be either a subset of the nonnegative integers $\{0, 1, 2, \ldots\}$ or a subset of $[0, \infty)$, the nonnegative real numbers. In the first case we will call the process *discrete time*, and in the second case *continuous time*. The random variables X_t will take values in a set that we call the *state space*. We will consider cases both where the state space is *discrete*, i.e., a finite or countably infinite set, and cases where the state space is *continuous*, e.g., the real numbers R or d-dimensional space R^d.

The study of deterministic (nonrandom) processes changing with time leads one to the study of differential equations (if time is continuous) or difference equations (if time is discrete). A typical (first-order) differential equation is of the form

$$y'(t) = F(t, y(t)).$$

Here the change in the function $y(t)$ depends only on t and the value $y(t)$ and not on the values at times before t. A large class of stochastic processes also have the property that the change at time t is determined by the value of the process at time t and not by the values at times before t. Such processes are called *Markov processes*. The study of such processes is closely related to linear algebra, differential equations, and difference equations. We assume that the reader is familiar with linear algebra. In the next section we review some facts about linear differential equations that will be used and in the following section we discuss difference equations.

0.2 Linear Differential Equations

Here we briefly review some facts about homogeneous linear differential equations with constant coefficients. Readers who want more detail should consult any introductory text in differential equations. Consider the homogeneous differential equation

$$y^{(n)}(t) + a_{n-1}y^{(n-1)}(t) + \cdots + a_1 y'(t) + a_0 y(t) = 0, \qquad (0.1)$$

where $a_0, \ldots, a_{n-1}$ are constants. For any initial conditions

$$y(0) = b_0, \ y'(0) = b_1, \ \ldots, \ y^{(n-1)}(0) = b_{n-1},$$

there is a unique solution to (0.1) satisfying these conditions. To obtain such a particular solution, we first find the general solution. Suppose $y_1(t), \ldots,$ $y_n(t)$ are linearly independent solutions to (0.1). Then every solution can be written in the form

$$y(t) = c_1 y_1(t) + \cdots + c_n y_n(t),$$

for constants $c_1, \ldots, c_n$. For a given set of initial conditions we can determine the appropriate constants.

The solutions $y_1, \ldots, y_n$ are found by looking for solutions of the form $y(t) = e^{\lambda t}$. Plugging in, we see that such a function $y(t)$ satisfies the equation if and only if

$$\lambda^n + a_{n-1}\lambda^{n-1} + \cdots + a_1\lambda + a_0 = 0.$$

If this polynomial has n distinct roots $\lambda_1, \ldots, \lambda_n$ we get n linearly independent solutions $e^{\lambda_1 t}, \ldots, e^{\lambda_n t}$. The case of repeated roots is a little trickier, but with a little calculation one can show that if λ is a root of multiplicity j, then $e^{\lambda t}, te^{\lambda t}, \ldots, t^{j-1}e^{\lambda t}$ are all solutions. Hence for each root of multiplicity j, we get j linearly independent solutions, and combining them all we get n linearly independent solutions as required.

Now consider the first-order linear system of equation

$$
\begin{aligned}
y_1'(t) &= a_{11}y_1(t) + a_{12}y_2(t) + \cdots + a_{1n}y_n(t) \\
y_2'(t) &= a_{21}y_1(t) + a_{22}y_2(t) + \cdots + a_{2n}y_n(t) \\
&\ \ \vdots \\
y_n'(t) &= a_{n1}y_1(t) + a_{n2}y_2(t) + \cdots + a_{nn}y_n(t).
\end{aligned}
$$

This can be written as a single vector valued equation:

$$\bar{y}'(t) = A\bar{y}(t).$$

Here $\bar{y}(t) = [y_1(t), \ldots, y_n(t)]$ (more precisely, the transpose of this vector) and A is the matrix of coefficients (a_{ij}). For any initial vector $\bar{v} = (v_1, \ldots, v_n)$, there is a unique solution to this equation satisfying $\bar{y}(0) = \bar{v}$. This solution can most easily be written in terms of the exponential of the matrix,

$$\bar{y}(t) = e^{tA}\bar{v}.$$

This exponential can be defined in terms of a power series:

$$e^{tA} = \sum_{j=0}^{\infty} \frac{(tA)^j}{j!}.$$

For computational purposes one generally tries to diagonalize the matrix A. Suppose that $A = Q^{-1}DQ$ for some diagonal matrix

$$D = \begin{bmatrix} d_1 & 0 & \cdots & 0 \\ 0 & d_2 & \cdots & 0 \\ \vdots & \vdots & \vdots & \vdots \\ 0 & 0 & \cdots & d_n \end{bmatrix}.$$

Then

$$e^{tA} = Q^{-1}e^{tD}Q = Q^{-1} \begin{bmatrix} e^{td_1} & 0 & \cdots & 0 \\ 0 & e^{td_2} & \cdots & 0 \\ \vdots & \vdots & \vdots & \vdots \\ 0 & 0 & \cdots & e^{td_n} \end{bmatrix} Q.$$

It is not true that every matrix can be diagonalized as above. However, every matrix A can be written as $Q^{-1}JQ$ where J is in Jordan canonical form. Taking exponentials of matrices in Jordan form is only slightly more difficult than taking exponentials of diagonal matrices. See a text on linear algebra for more details.

0.3 Linear Difference Equations

The theory of linear difference equations is very similar to that of linear differential equations. However, since the theory is generally not studied in introductory differential equations courses and since difference equations arise naturally in discrete-time Markov chains, we will discuss their solution in more detail. First consider the equation

$$f(n) = af(n-1) + bf(n+1), \quad K < n < N. \tag{0.2}$$

Here $f(n)$ is a function defined for integers $K \leq n \leq N$ (N can be chosen to be infinity) and a, b are nonzero real numbers. If f satisfies (0.2) and the values $f(K)$ and $f(K+1)$ are known, then $f(n)$ can be determined for all $K \leq n \leq N$ recursively by the formula

$$f(n+1) = \frac{1}{b}[f(n) - af(n-1)]. \tag{0.3}$$

Conversely, if u_0, u_1 are any real numbers we can find a solution to (0.2) satisfying $f(K) = u_0, f(K+1) = u_1$ by defining $f(n)$ recursively as in (0.3). Also, we note that the set of functions satisfying (0.2) is a vector space, i.e., if f_1, f_2 satisfy (0.2) then so does $c_1 f_1 + c_2 f_2$, where c_1, c_2 are any real numbers. This vector space has dimension 2; in fact, a basis for the vector space is given by $\{f_1, f_2\}$, where f_1 is the solution satisfying $f_1(K) = 1, f_1(K+1) = 0$ and f_2 is the solution satisfying $f_2(K) = 0, f_2(K+1) = 1$. If g_1 and g_2 are any two linearly independent solutions, then it is a standard fact from linear algebra that every solution is of the form

$$c_1 g_1 + c_2 g_2$$

for constants c_1, c_2.

We now make some good guesses to find a pair of linearly independent solutions. We will try functions of the form $f(n) = \alpha^n$ for some $\alpha \neq 0$. This is a solution for a particular α if and only if

$$\alpha^n = a\alpha^{n-1} + b\alpha^{n+1}, \quad K < n < N,$$

i.e., if and only if

$$\alpha = a + b\alpha^2.$$

We can solve this with the quadratic formula, giving

$$\alpha = \frac{1 \pm \sqrt{1 - 4ab}}{2b}.$$

Case I: $1 - 4ab \neq 0$. In this case there are two distinct roots, α_1, α_2, and hence the general solution is

$$f(n) = c_1 \alpha_1^n + c_2 \alpha_2^n. \tag{0.4}$$

Case II: $1 - 4ab = 0$. In this case we get only one solution of this type, $g_1(n) = \alpha^n = (1/2b)^n$. However, if we let $g_2(n) = n(1/2b)^n$ we see that

$$\begin{aligned} ag_2(n-1) + bg_2(n+1) &= a(n-1)(1/2b)^{n-1} + b(n+1)(1/2b)^{n+1} \\ &= (1/2b)^n[a(n-1)2b + b(n+1)/(2b)] \\ &= (1/2b)^n n = g_2(n). \end{aligned}$$

Therefore g_2 is also a solution. It is easy to check that g_1, g_2 are linearly independent, so every solution is of the form

$$f(n) = c_1(1/2b)^n + c_2 n(1/2b)^n.$$

Example. Suppose we want to find a function f satisfying

$$f(n) = (1/6)f(n-1) + (2/3)f(n+1), \quad 0 < n < \infty,$$

with $f(0) = 4, f(1) = 3$. Plugging in we get,

$$\alpha = \frac{3 \pm \sqrt{5}}{4}.$$

The general solution is

$$f(n) = c_1\left(\frac{3+\sqrt{5}}{4}\right)^n + c_2\left(\frac{3-\sqrt{5}}{4}\right)^n.$$

If we plug in the initial conditions, we get

$$4 = f(0) = c_1 + c_2,$$

$$3 = f(1) = c_1 \frac{3 + \sqrt{5}}{4} + c_2 \frac{3 - \sqrt{5}}{4}.$$

Solving gives $c_1 = 2, c_2 = 2$, and hence

$$f(n) = 2(\frac{3 + \sqrt{5}}{4})^n + 2(\frac{3 - \sqrt{5}}{4})^n.$$

We have seen that the values of $f(K)$ and $f(K + 1)$ uniquely determine the solution to (0.2). Sometimes, one is given the boundary values $f(K)$ and $f(N)$. These boundary value problems can be solved in the same way— write down the general solution and solve for the constants. For example, suppose we want the function f which satisfies

$$f(n) = 2f(n - 1) - f(n + 1), \quad 0 < n < 10,$$

with $f(0) = 0, f(10) = 1$. We write down the general solution

$$f(n) = c_1 1^n + c_2(-2)^n.$$

Plugging in the initial conditions gives

$$f(0) = 0 = c_1 + c_2$$

$$f(10) = 1 = c_1 + c_2(-2)^n,$$

and $c_1 = -c_2 = 1/(1 - 2^{10})$.

In the study of random walks, the difference equations

$$f(n) = (1 - p)f(n - 1) + pf(n + 1), \quad p \in (0, 1)$$

arise. If $p \neq 1/2$, we obtain two roots $\alpha_1 = 1, \alpha_2 = (1 - p)/p$, and hence the general solution is

$$f(n) = c_1 + c_2(\frac{1 - p}{p})^n. \tag{0.5}$$

If $p = 1/2$, $\alpha = 1$ is a repeated root so we get the general solution

$$f(n) = c_1 + c_2 n. \tag{0.6}$$

What we have analyzed are second-order linear difference equations. The general kth-order homogeneous linear difference equation is of the form

$$f(n + k) = a_0 f(n) + a_1 f(n + 1) + \cdots + a_{k-1} f(n + k - 1). \tag{0.7}$$

Suppose we wish to find a function satisfying (0.7) for $n \geq 0$. It suffices to give the values $f(0), \ldots, f(k - 1)$, for then $f(n), n \geq k$ can be determined recursively. Again we look for solutions of the form $f(n) = \alpha^n$. Such an f is a solution if and only if

$$\alpha^k = a_0 + a_1 \alpha + \cdots + a_{k-1} \alpha^{k-1}.$$

As before, if there are k distinct roots of this equation, we get k linearly independent solutions. If a certain α is a root with multiplicity j, one can check in fact that

$$\alpha^n, n\alpha^n, n^2\alpha^n, \cdots, n^{j-1}\alpha^n$$

are all linearly independent solutions. In complete parallel with the case of linear differential equations, we get k linearly independent solutions to (0.7) and we can find all solutions by taking linearly combinations of these solutions.

0.4 Exercises

0.1 Find all functions $x(t), y(t)$ satisfying

$$x'(t) = y(t) - x(t),$$

$$y'(t) = 3x(t) - 3y(t).$$

Find the particular pair of functions satisfying $x(0) = y(0) = 1/2$.

0.2 Find the function $f(n), n = 0, 1, \ldots, 10$ that satisfies

$$f(n) = \frac{1}{4}f(n-1) + \frac{3}{4}f(n+1), \quad n = 1, 2, \ldots, 9,$$

and $f(0) = 0, f(1) = 1$.

0.3 The Fibonacci numbers F_n are defined by $F_1 = 1$, $F_2 = 1$ and for $n > 2$, $F_n = F_{n-1} + F_{n-2}$. Find a formula for F_n by solving the difference equation.

0.4 Find the function $f(n)$, $n = 0, 1, 2, \ldots$ that satsifies

$$f(0) = 0,$$

$$f(n) = \frac{1}{3}f(n-1) + \frac{1}{3}f(n+1) + \frac{1}{3}f(n+2), \quad n \geq 1,$$

$$\lim_{n \to \infty} f(n) = 1.$$

0.5 Find all functions f from the integers to the real numbers satisfying

$$f(n) = \frac{1}{2}f(n+1) + \frac{1}{2}f(n-1) - 1. \tag{0.8}$$

[Hint: First show that $f(n) = n^2$ satisfies (0.8). Then suppose $f_1(n)$ and $f_2(n)$ both satisfy (0.8) and find the equation that $g(n) = f_2(n) - f_1(n)$ satisfies.]

CHAPTER 1

Finite Markov Chains

1.1 Definitions and Examples

Consider a discrete-time stochastic process, $X_n, n = 0, 1, 2, \ldots$, where X_n takes values in the finite set $S = \{1, \ldots, N\}$ or $\{0, \ldots, N-1\}$. We call the possible values for X_n the *states* of the system. To describe the probabilities for such a process we need to give the values of

$$P\{X_0 = i_0, X_1 = i_1, \ldots, X_n = i_n\},$$

for every n and every finite sequence of states $(i_0, \ldots, i_n)$. Equivalently, we could give the initial probability distribution

$$\phi(i) = P\{X_0 = i\}, \quad i = 1, \ldots, N$$

and the "transition probabilities",

$$q_n(i_n \mid i_0, \ldots, i_{n-1}) = P\{X_n = i_n \mid X_0 = i_0, \ldots, X_{n-1} = i_{n-1}\}, \quad (1.1)$$

for then

$$P\{X_0 = i_0, \ldots, X_n = i_n\} =$$

$$\phi(i_0) q_1(i_1 \mid i_0) q_2(i_2 \mid i_0, i_1) \cdots q_n(i_n \mid i_0, \ldots, i_{n-1}). \quad (1.2)$$

In this chapter we consider a special class of such processes, those that satisfy the *Markov property*. The Markov property states that to make predictions of the behavior of a system in the future, it suffices to consider only the present state of the system and not the past history. That is, the state of the system is important but not how it arrived at that state. Mathematically, we can write this as

$$P\{X_n = i_n \mid X_0 = i_0, \ldots, X_{n-1} = i_{n-1}\} = P\{X_n = i_n \mid X_{n-1} = i_{n-1}\}.$$

We will also make another assumption: the transition probabilities do not depend on time. This is called time homogeneity. A *time-homogeneous Markov chain* is a process such that

$$P\{X_n = i_n \mid X_0 = i_0, \ldots, X_{n-1} = i_{n-1}\} = p(i_{n-1}, i_n),$$

for some function $p : S \times S \to [0, 1]$. Unless explicitly stated otherwise in this book, when we say Markov chain we will mean time-homogeneous

Markov chain. To give the probabilities for a Markov chain, we need to give an initial probability distribution $\phi(i) = P\{X_0 = i\}$, and the transition probabilities $p(i, j)$, for then, by (1.2),

$$P\{X_0 = i_0, \ldots, X_n = i_n\} = \phi(i_0)p(i_0, i_1)p(i_1, i_2) \cdots p(i_{n-1}, i_n). \quad (1.3)$$

The *transition matrix* P for the Markov chain is the $N \times N$ matrix whose (i, j) entry P_{ij} is $p(i, j)$. The matrix P is a *stochastic matrix*, i.e.,

$$0 \leq P_{ij} \leq 1, \quad 1 \leq i, j \leq N, \quad (1.4)$$

$$\sum_{j=1}^{N} P_{ij} = 1, \quad 1 \leq i \leq N. \quad (1.5)$$

Any matrix satisfying (1.4) and (1.5) can be the transition matrix for a Markov chain.

Example 1. Two-state Markov chain. Consider the state of a phone where $X_n = 0$ means that the phone is free at time n and $X_n = 1$ means that the phone is busy. We assume that during each time interval there is a probability p that a call comes in (for ease we will assume that no more than one call comes in during any particular time interval). If the phone is busy during that period, the incoming call does not get through. We also assume that if the phone is busy during a time interval, there is a probability q that it will be free during the next interval. This gives a Markov chain with state space $S = \{0, 1\}$ and matrix

$$P = \begin{bmatrix} 1 - p & p \\ q & 1 - q \end{bmatrix}.$$

This matrix give the general form for a transition matrix of a two-state Markov chain. In order to specify the matrix one only needs to give the values of p and q.

Example 2. Simple Queueing Model. We adapt the previous example by assuming that the phone system can put one caller on hold. Hence at any time the number of callers in the system is in the set $S = \{0, 1, 2\}$. Again, any call will be completed during a time interval with probability q and a new caller will come in with probability p, unless the system is already full. To model this we set

$$p(0, 0) = 1 - p, \quad p(0, 1) = p, \quad p(0, 2) = 0,$$

since a caller comes in with probability p (again we are assuming only one caller arrives during any time period). Also,

$$p(2, 0) = 0, \quad p(2, 1) = q, \quad p(2, 2) = 1 - q,$$

since no new callers may arrive if there are two callers in the system, and both calls may not end simultaneously. If there is exactly one caller in the system, it is a little more complicated. The state of the system goes from 1 to 0 if the current call is completed and no new callers enter the system, i.e., $p(1,0) = q(1-p)$. Similarly, the state goes from 1 to 2 if the current call is not completed but a new call arrives, i.e., $p(1,2) = p(1-q)$. Since the rows must add to 1, $p(1,1) = 1 - q(1-p) - p(1-q)$ and hence

$$P = \begin{bmatrix} 1-p & p & 0 \\ q(1-p) & 1 - q(1-p) - p(1-q) & p(1-q) \\ 0 & q & 1-q \end{bmatrix}.$$

Transition probabilities are often represented by directed graphs, where the vertices of the graphs are the states and the arrows represent the transitions. The above matrix can be represented graphically as follows:

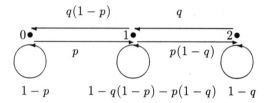

Example 3. Random Walk with Reflecting Boundary. Consider a "random walker" moving along the sites $\{0, 1, \ldots, N\}$.

At each time step the walker moves one step, to the right with probability p and to the left with probability $1 - p$. If the walker is at one of the boundary points $\{0, N\}$, the walker moves with probability 1 toward the inside of the interval. The transition matrix P for this Markov chain is given by

$$p(i, i+1) = p, \quad p(i, i-1) = 1-p, \quad 0 < i < N,$$

$$p(0, 1) = 1, \quad p(N, N-1) = 1,$$

with $p(i, j) = 0$ for other values of i, j. If $p = 1/2$, we call this *symmetric* or *unbiased* random walk with reflecting boundaries. If $p \neq 1/2$ it is called *biased* random walk. Sometimes it is more convenient to consider partially reflecting boundaries where the walker at the boundary moves the same as on the inside except that if the walker tries to leave the states $\{0, \ldots, N\}$ he runs into a wall and goes nowhere. This corresponds to boundary conditions

$$p(0,0) = 1 - p, \ p(0,1) = p; \quad p(N, N-1) = 1 - p, \ p(N, N) = p.$$

Example 4. Random Walk with Absorbing Boundaries. This chain is like the previous example except that when the walker reaches 0 or N, the walker stays there forever. The transition matrix is given by

$$p(i, i+1) = p, \quad p(i, i-1) = 1 - p, \quad 0 < i < N,$$

$$p(0,0) = 1, \quad p(N, N) = 1.$$

(We adopt the convention from here on that if $p(i, j)$ is not specified for a particular i, j then it is assumed to be 0.)

Example 5. Simple Random Walk on a Graph. A *(finite, simple, undirected) graph* is a finite collection of vertices V and a collection of edges E where each edge connects two different vertices and any two vertices are connected by at most one edge. We write $v_1 \sim v_2$ if vertices v_1 and v_2 are adjacent, i.e., an edge connects the two vertices.

Consider the Markov chain whose states are the vertices of the graph. At each time interval, the chain chooses a new state randomly from among the states adjacent to the current state. The transition matrix for this chain is given by

$$p(v_i, v_j) = 1/d(v_i), \quad v_i \sim v_j,$$

where $d(v_i)$ is the number of vertices adjacent to v_i [if $d(v_i) = 0$, we let $p(v_i, v_i) = 1.$]. This chain is called *simple random walk on the graph.* Symmetric random walk ($p = 1/2$) with reflecting boundaries as in Example 3 is a particular example of a simple random walk on a graph.

Given a transition matrix P and an initial probability distribution ϕ, how can we determine the probabilities that the Markov chain will be in a certain state i at a given time n? We define the n-step probabilities $p_n(i, j)$

by

$$p_n(i,j) = P\{X_n = j \mid X_0 = i\} = P\{X_{n+k} = j \mid X_k = i\}$$

(the latter equality holds because of time homogeneity). Then

$$P\{X_n = j\} = \sum_{i \in S} \phi(i) P\{X_n = j \mid X_0 = i\}. \qquad (1.6)$$

We will now show that the n-step transition probability is in fact the (i,j) entry in the matrix P^n. To see this, we first note that this is trivially true for $n = 1$. Assume it is true for a given n. Then,

$$
\begin{aligned}
&P\{X_{n+1} = j \mid X_0 = i\} \\
&= \sum_k P\{X_n = k \mid X_0 = i\} P\{X_{n+1} = j \mid X_n = k\} \\
&= \sum_k p_n(i,k) p(k,j).
\end{aligned}
$$

But if $p_n(i,k)$ is the (i,k) entry of P^n, the last sum is exactly the (i,j) entry of $P^n P = P^{n+1}$.

An initial probability distribution can be given by a vector

$$\bar{\phi}_0 = (\phi_0(1), \ldots, \phi_0(N)).$$

[We will denote the vector $(v(1), \ldots, v(N))$ by $\bar{v}$. We will use the same notation whether $\bar{v}$ is to be considered a row vector or a column vector. For example, we can write either $\bar{v}P$, or $P\bar{v}$ although $\bar{v}$ is a row vector in the first case and a column vector in the second.] If $\bar{\phi}_0$ is given, the distribution at time n, $\phi_n(i) = P\{X_n = i\}$ is given by

$$\bar{\phi}_n = \bar{\phi}_0 P^n.$$

Example 6. Consider Example 1 and assume the phone is free at time 0. Assume $p = 1/4$ and $q = 1/6$. Let $n = 6$. Then

$$P^6 = \begin{bmatrix} 3/4 & 1/4 \\ 1/6 & 5/6 \end{bmatrix}^6 = \begin{bmatrix} .424 & .576 \\ .384 & .616 \end{bmatrix}.$$

If the phone is free at time 0, $\bar{\phi}_0 = (1,0)$. If we want to know the probability that the phone is busy at time 6, we compute

$$(\bar{\phi}_0 P^6)(1) = .576.$$

1.2 Long-Range Behavior and Invariant Probability

Understanding the long-range behavior of a Markov chain boils down to understanding the behavior of P^n for large n values. Let us start by considering a particular example,

$$P = \begin{bmatrix} 3/4 & 1/4 \\ 1/6 & 5/6 \end{bmatrix}.$$

Taking powers of this matrix is easy (with a computer) and one can quickly see that

$$P^n \approx \begin{bmatrix} .4 & .6 \\ .4 & .6 \end{bmatrix},$$

for large n, i.e., a limit matrix

$$\Pi = \lim_{n \to \infty} P^n$$

exists and the rows of Π are identical. If $\bar{v}$ is any probability vector [we say a vector $\bar{v} = (v(1), \ldots, v(N))$ is a probability vector if the components are nonnegative and sum to 1], then

$$\lim_{n \to \infty} \bar{v} P^n = \bar{\pi},$$

where $\bar{\pi} = (2/5, 3/5)$ is one of the rows of Π. For another example, consider Example 2 of Section 1.1 with $p = 1/4, q = 1/6$,

$$P = \begin{bmatrix} 3/4 & 1/4 & 0 \\ 1/8 & 2/3 & 5/24 \\ 0 & 1/6 & 5/6 \end{bmatrix}. \tag{1.7}$$

We see the same phenomenon. In this case for large n,

$$P^n \approx \begin{bmatrix} .182 & .364 & .455 \\ .182 & .364 & .455 \\ .182 & .364 & .455 \end{bmatrix} = \begin{bmatrix} \bar{\pi} \\ \bar{\pi} \\ \bar{\pi} \end{bmatrix},$$

where $\bar{\pi} = (2/11, 4/11, 5/11)$ and hence for every $\bar{v}$,

$$\lim_{n \to \infty} \bar{v} P^n = \bar{\pi}.$$

At any large time, the probability that the phone has no callers is about $\pi(0) = 2/11$, regardless of what the state of the system was at time 0.

Suppose $\bar{\pi}$ is a limiting probability vector, i.e., for some initial probability vector $\bar{v}$,

$$\bar{\pi} = \lim_{n \to \infty} \bar{v} P^n.$$

Then

$$\bar{\pi} = \lim_{n \to \infty} \bar{v} P^{n+1} = (\lim_{n \to \infty} \bar{v} P^n) P = \bar{\pi} P.$$

We call a probability vector $\bar{\pi}$ an *invariant probability distribution* for P if

$$\bar{\pi} = \bar{\pi} P. \tag{1.8}$$

Such a $\bar{\pi}$ is also called a *stationary, equilibrium,* or *steady-state* probability distribution. Note that an invariant probability vector is a left eigenvector of P with eigenvalue 1.

There are three natural questions to ask about invariant probability distributions for stochastic matrices:

1) Does every stochastic matrix P have an invariant probability distribution $\bar{\pi}$?

2) Is the invariant probability distribution unique?

3) When can we conclude that

$$\lim_{n \to \infty} P^n = \begin{bmatrix} \bar{\pi} \\ \bar{\pi} \\ \vdots \\ \bar{\pi} \end{bmatrix},$$

and hence that for all initial probability distributions $\bar{v}$,

$$\lim_{n \to \infty} \bar{v} P^n = \bar{\pi}?$$

Let us start by considering the two-state Markov chain with

$$P = \begin{bmatrix} 1-p & p \\ q & 1-q \end{bmatrix},$$

where $0 < p, q < 1$. This matrix has eigenvalues 1 and $1 - p - q$. We can diagonalize the matrix P,

$$D = Q^{-1} P Q,$$

where

$$Q = \begin{bmatrix} 1 & -p \\ 1 & q \end{bmatrix}, \quad Q^{-1} = \begin{bmatrix} q/(p+q) & p/(p+q) \\ -1/(p+q) & 1/(p+q) \end{bmatrix}.$$

$$D = \begin{bmatrix} 1 & 0 \\ 0 & 1-p-q \end{bmatrix}.$$

The columns of Q are right eigenvectors of P and the rows of Q^{-1} are left eigenvectors. The eigenvectors are unique up to a multiplicative constant. We have chosen the constant in the left eigenvector for eigenvalue 1 so that it is a probability vector. $\bar{\pi} = (q/(p+q), p/(p+q))$ is the unique invariant probability distribution for P. Once P is diagonalized it is easy to raise P to powers,

$$\begin{aligned} P^n &= (QDQ^{-1})^n \\ &= QD^nQ^{-1} \\ &= Q \begin{bmatrix} 1 & 0 \\ 0 & (1-p-q)^n \end{bmatrix} Q^{-1} \\ &= \begin{bmatrix} [q + p(1-p-q)^n]/(p+q) & [p - p(1-p-q)^n]/(p+q) \\ [q - q(1-p-q)^n]/(p+q) & [p + q(1-p-q)^n]/(p+q) \end{bmatrix}. \end{aligned}$$

Since $|1 - p - q| < 1$, we see that

$$\lim_{n \to \infty} P^n = \begin{bmatrix} q/(p+q) & p/(p+q) \\ q/(p+q) & p/(p+q) \end{bmatrix} = \begin{bmatrix} \bar{\pi} \\ \bar{\pi} \end{bmatrix}.$$

The key to the computation of the limit is the fact that the second eigen-value $1 - p - q$ has absolute value less than 1 and so the dominant contribu-tion to P^n comes from the eigenvector with eigenvalue 1, i.e., the invariant probability distribution.

Suppose P is any stochastic matrix. It is easy to check that the vector $\bar{1} = (1, 1, \cdots, 1)$ is a *right* eigenvector with eigenvalue 1. Hence at least one left eigenvector for eigenvalue 1 exists. Suppose we can show that:

The left eigenvector can be chosen to have all nonnegative entries,

(1.9)

The eigenvalue 1 is simple and all other eigenvalues

have absolute value less than 1.   (1.10)

Then we can show that essentially the same thing happens as in the two-state case. It is not always true that we can diagonalize P; however, we can do well enough using a Jordon decomposition (consult a text in linear algebra for details): there exists a matrix Q such that

$$D = Q^{-1}PQ,$$

where the first row of Q^{-1} is the unique invariant probability vector $\bar{\pi}$; the first column of Q contains all 1s. The matrix D is not necessarily diagonal but it does have the form

$$D = \begin{bmatrix} 1 & 0 & \cdots & 0 \\ 0 & & & \\ \vdots & & M & \\ 0 & & & \end{bmatrix},$$

where $M^n \to 0$. Then in the same way as the two-state example,

$$\lim_{n \to \infty} P^n = \lim_{n \to \infty} QD^nQ^{-1} = Q \begin{bmatrix} 1 & 0 & \cdots & 0 \\ 0 & & & \\ \vdots & & 0 & \\ 0 & & & \end{bmatrix} Q^{-1} = \begin{bmatrix} \bar{\pi} \\ \vdots \\ \bar{\pi} \end{bmatrix}.$$

This leads us to ask which matrices satisfy (1.9) and (1.10). The Perron–Frobenius Theorem from linear algebra gives one large class of matrices for which this is true. Suppose that P is a stochastic matrix such that all of the entries are strictly positive. Then the Perron–Frobenius Theorem implies that: 1 is a simple eigenvalue for P; the left eigenvector of 1 can be chosen to have all positive entries (and hence can be made into a probability vector by multiplying by an appropriate constant); and all the other eigenvalues have absolute value strictly less than 1. We sketch a proof of this theorem in Exercise 1.15.

While this includes a large number of matrices, it does not cover all stochastic matrices with the appropriate limit behavior. For example, consider the matrix P in (1.7). While this does not have all positive entries note that

$$P^2 = \begin{bmatrix} .594 & .354 & .052 \\ .177 & .510 & .312 \\ .021 & .250 & .729 \end{bmatrix},$$

and hence P^2 satisfies the conditions of the theorem. Therefore, 1 is a simple eigenvalue for P^2 with invariant probability $\bar{\pi}$ and the other eigenvalues of P^2 have absolute value strictly less than 1. Since the eigenvalues for P^2 are the squares of the eigenvalues of P, and eigenvectors of P are eigenvectors of P^2, we see that P also satisfies (1.9) and (1.10). We then get a general rule.

Rule. *If P is a stochastic matrix such that for some n, P^n has all entries strictly positive, then P satisfies (1.9) and (1.10).*

In the next section we discuss which matrices P have the property that P^n has all positive entries for some n.

1.3 Classification of States

In this section we investigate under what conditions on a stochastic matrix P we can conclude that P^n has all positive entries for some sufficiently large n. We start by considering some examples where this is not true.

Example 1. Simple random walk with reflecting boundary on $\{0,\ldots,4\}$. In this case,

$$P = \begin{bmatrix} 0 & 1 & 0 & 0 & 0 \\ 1/2 & 0 & 1/2 & 0 & 0 \\ 0 & 1/2 & 0 & 1/2 & 0 \\ 0 & 0 & 1/2 & 0 & 1/2 \\ 0 & 0 & 0 & 1 & 0 \end{bmatrix}.$$

If one takes powers of this matrix, one quickly sees that P^n looks different depending on whether n is even or odd. For large n, if n is even,

$$P^n \approx \begin{bmatrix} .25 & 0 & .50 & 0 & .25 \\ 0 & .50 & 0 & .50 & 0 \\ .25 & 0 & .50 & 0 & .25 \\ 0 & .50 & 0 & .50 & 0 \\ .25 & 0 & .50 & 0 & .25 \end{bmatrix},$$

whereas if n is odd,

$$
P^n \approx
\begin{bmatrix}
0 & .50 & 0 & .50 & 0 \\
.25 & 0 & .50 & 0 & .25 \\
0 & .50 & 0 & .50 & 0 \\
.25 & 0 & .50 & 0 & .25 \\
0 & .50 & 0 & .50 & 0
\end{bmatrix}.
$$

It is easy to see why there should be many zeroes in P^n. At each step, the random walker moves from an "even" step to an "odd" step or vice versa. If the walker starts on an even site, then after an even number of steps the walker will be on an even site, i.e., $p_n(i,j) = 0$ if i is even, j is odd, n is even. Similarly, after an odd number of steps, a walker who started on an even point will be at an odd point. In this example we say that P has period 2.

Example 2. Simple random walk with absorbing boundary on $\{0, \ldots, 4\}$. Here,

$$
P =
\begin{bmatrix}
1 & 0 & 0 & 0 & 0 \\
1/2 & 0 & 1/2 & 0 & 0 \\
0 & 1/2 & 0 & 1/2 & 0 \\
0 & 0 & 1/2 & 0 & 1/2 \\
0 & 0 & 0 & 0 & 1
\end{bmatrix}.
$$

If n is large, we see that

$$
P^n \approx
\begin{bmatrix}
1 & 0 & 0 & 0 & 0 \\
.75 & 0 & 0 & 0 & .25 \\
.50 & 0 & 0 & 0 & .50 \\
.25 & 0 & 0 & 0 & .75 \\
0 & 0 & 0 & 0 & 1
\end{bmatrix}.
$$

In this case the random walker eventually gets to 0 or 4 and then stays at that state forever. Consider the second row. Note that $p_n(1,0) \to 3/4$ and $p_n(1,4) \to 1/4$. This implies that the probability that a random walker starting at 1 will eventually stick at 0 is 3/4, whereas with probability 1/4 she eventually sticks at 4. We will call states such as $1, 2, 3$ transient states of the Markov chain.

Example 3. Suppose $S = \{1, 2, 3, 4, 5\}$ and

$$
P \Rightarrow
\begin{bmatrix}
1/2 & 1/2 & 0 & 0 & 0 \\
1/6 & 5/6 & 0 & 0 & 0 \\
0 & 0 & 3/4 & 1/4 & 0 \\
0 & 0 & 1/8 & 2/3 & 5/24 \\
0 & 0 & 0 & 1/6 & 5/6
\end{bmatrix}.
$$

For large n,

$$P^n \approx \begin{bmatrix} .25 & .75 & 0 & 0 & 0 \\ .25 & .75 & 0 & 0 & 0 \\ 0 & 0 & .182 & .364 & .455 \\ 0 & 0 & .182 & .364 & .455 \\ 0 & 0 & .182 & .364 & .455 \end{bmatrix}.$$

In this case the chain splits into two smaller, noninteracting chains: a chain with state space $\{1,2\}$ and a chain with state space $\{3,4,5\}$. Each "sub-chain" converges to an equilibrium distribution, but one cannot change from a state in $\{1,2\}$ to a state in $\{3,4,5\}$. We call such a system a reducible Markov chain.

The main goal of this section is to show that the above examples illustrate all the ways that a Markov chain can fail to satisfy (1.9) and (1.10).

Reducibility.

We say two states i and j of a Markov chain *communicate* with each other, written $i \leftrightarrow j$, if there exist $m, n \geq 0$ such that $p_m(i,j) > 0$ and $p_n(j,i) > 0$. In other words, two states communicate if and only if each state has a positive probability of eventually being reached by a chain starting in the other state. The relation $\leftrightarrow$ is an equivalence relation on the state space, i.e., it is: *reflexive*, $i \leftrightarrow i$ [since $p_0(i,i) = 1 > 0$]; *symmetric*, $i \leftrightarrow j$ implies that $j \leftrightarrow i$ (this is immediate from the definition); and *transitive*, $i \leftrightarrow j$ and $j \leftrightarrow k$ imply $i \leftrightarrow k$ [to see this note that if $p_{m_1}(i,j) > 0$ and $p_{m_2}(j,k) > 0$ then

$$\begin{aligned} p_{m_1+m_2}(i,k) &= P\{X_{m_1+m_2} = k \mid X_0 = i\} \\ &\geq P\{X_{m_1+m_2} = k, X_{m_1} = j \mid X_0 = i\} \\ &= P\{X_{m_1} = j \mid X_0 = i\}P\{X_{m_1+m_2} = k \mid X_{m_1} = j\} \\ &= p_{m_1}(i,j)p_{m_2}(j,k) > 0, \end{aligned}$$

and similarly $p_{n_1}(j,i) > 0, p_{n_2}(k,j) > 0$ imply $p_{n_1+n_2}(k,i) > 0$]. This equivalence relation partitions the state space into disjoint sets called *communication classes*. For example, in Example 3 in this section there are two communication classes $\{1,2\}$ and $\{3,4,5\}$.

If there is only one communication class, i.e., if for all i,j there exists an $n = n(i,j)$ with $p_n(i,j) > 0$, then the chain is called *irreducible*. Any matrix satisfying (1.9) and (1.10) is irreducible. However, one can also check that Example 1 of this section is also irreducible. Example 2 has three communication classes, $\{0\}, \{1,2,3\}$, and $\{4\}$. In this example, if the chain starts in the class $\{1,2,3\}$, then with probability 1 it eventually leaves this class and never returns. Classes with this property are called *transient* classes and the states are called transient states. Other classes are called *recurrent* classes with recurrent states. A Markov chain starting in a recurrent class never leaves that class.

Suppose P is the matrix for a reducible Markov chain with recurrent communication classes $R_1, \ldots, R_r$ and transient classes $T_1, \ldots, T_s$. It is easy to see that there must be at least one recurrent class. For each recurrent class R, the submatrix of P obtained from considering only the rows and columns for states in R is a stochastic matrix. Hence we can write P in the following form (after, perhaps, reordering the states):

$$
P = \left[
\begin{array}{ccccc|c}
P_1 & & & & & \\
& P_2 & & & 0 & \\
& & P_3 & & & 0 \\
& 0 & & \ddots & & \\
& & & & P_r & \\
\hline
& & S & & & Q
\end{array}
\right],
$$

where P_k is the matrix associated with R_k. Then,

$$
P^n = \left[
\begin{array}{ccccc|c}
P_1^n & & & & & \\
& P_2^n & & & 0 & \\
& & P_3^n & & & 0 \\
& 0 & & \ddots & & \\
& & & & P_r^n & \\
\hline
& & S_n & & & Q^n
\end{array}
\right],
$$

for some matrix S_n. To analyze the large time behavior of the Markov chain on the class R_k we need only consider the matrix P_k. We discuss the behavior of Q^n in Section 1.5.

Periodicity.

Suppose that P is the matrix for an irreducible Markov chain (if P is reducible we can consider separately each of the recurrent communication classes). We define the *period* of a state i, $d = d(i)$, to be the greatest common divisor of

$$J_i \doteq \{n \geq 0 : p_n(i, i) > 0\}.$$

In Example 1 of this section, the period of each state is 2; in fact, in this case $p_{2n}(i, i) > 0$ and $p_{2n+1}(i, i) = 0$ for all n, i.

Let J be any nonempty subset of the nonnegative integers that is closed under addition, i.e., $m, n \in J \Rightarrow m + n \in J$. An example of such a J is the set J_i since $p_{m+n}(i, i) \geq p_m(i, i) p_n(i, i)$. Let d be the greatest common divisor of the elements of J. Then $J \subset \{0, d, 2d, \ldots\}$. Moreover it can be shown (Exercise 1.16) that J must contain all but a finite number of the elements of $\{0, d, 2d, \ldots\}$, i.e., there is some M such that $md \in J$ for all $m > M$. Hence J_i contains md for all m greater than some $M = M_i$. If j is another state and m, n are such that $p_m(i, j) > 0, p_n(j, i) > 0$, then $m + n \in J_i, m + n \in J_j$. Hence $m + n = kd$ for some integer k. Also, if $l \in J_j$, then

$$p_{m+n+l}(i, j) \geq p_m(i, j)p_l(j, j)p_n(j, i) > 0,$$

and so d divides l. We have just shown that if d divides every element of J_i then it divides every element of J_j. From this we see that all states have the same period and hence we can talk about the period of P. (We have used the fact that P is irreducible. If P is reducible, it is possible for states in different communication classes to have different periods.)

Example 4. Consider simple random walk on a graph (see Example 5, Section 1.1). The chain is irreducible if and only if the graph is connected, i.e., if any two vertices can be connected by a path of edges in the graph. Every vertex in a connected graph (with at least two vertices) is adjacent to at least one other point. If $v \sim w$ then $p_2(v, v) \geq p_1(v, w)p_1(w, v) > 0$. Therefore, the period is either 1 or 2. It is easy to see that the period is 2 if and only if the graph is bipartite, i.e, if and only if the vertices can be partitioned into two disjoint sets V_1, V_2 such that all edges of the graph connect one vertex of V_1 and one vertex V_2. Note that symmetric random walk with reflecting boundaries gives an example of simple random walk on a bipartite graph.

We call an irreducible matrix P *aperiodic* if $d = 1$. What we will show now is the following: if P is irreducible and aperiodic, then there exist an $M > 0$ such that for all $n \geq M$, P^n has all entries strictly positive. To see this, take any i, j. Since P is irreducible there exists some $m(i, j)$ such that $p_{m(i,j)}(i, j) > 0$. Moreover, since P is aperiodic, there exists some $M(i)$ such that for all $n \geq M(i), p_n(i, i) > 0$. Hence for all $n \geq M(i)$,

$$p_{n+m(i,j)}(i, j) \geq p_n(i, i)p_{m(i,j)}(i, j) > 0.$$

Let M be the maximum value of $M(i) + m(i, j)$ over all pairs (i, j) (the maximum exists since the state space is finite). Then $p_n(i, j) > 0$ for all $n \geq M$ and all i, j. Using the rule at the end of the last section we can now summarize with the following theorem.

Theorem. *If P is the transition matrix for an irreducible, aperiodic Markov chain, then there exists a unique invariant probability vector $\bar{\pi}$ satisfying*

$$\bar{\pi}P = \bar{\pi}.$$

If $\bar{\phi}$ is any initial probability vector,

$$\lim_{n \to \infty} \bar{\phi}P^n = \bar{\pi}.$$

Moreover, $\pi(i) > 0$ for each i.

We finish this section by discussing how P^n behaves when P is not irreducible and aperiodic. First, assume P is reducible with recurrent classes $R_1, \ldots, R_r$ and transient classes $T_1, \ldots, T_s$. Each recurrent class acts as a small Markov chain; hence, there exists r different invariant probability vectors $\bar{\pi}^1, \ldots, \bar{\pi}^r$ with $\bar{\pi}^k$ concentrated on R_k ($\pi^k(i) = 0$ if $i \notin R_k$). In other words, the eigenvalue 1 has multiplicity r with one eigenvector for each recurrent class. Assume, for ease, that the submatrix P_k for each recurrent class is aperiodic. Then if $i \in R_k$,

$$\lim_{n \to \infty} p_n(i,j) = \pi^k(j), \quad j \in R_k,$$

$$p_n(i,j) = 0, \quad j \notin R_k.$$

If i is any transient state, then the chain starting at i eventually ends up in a recurrent state. This means that for each transient state j,

$$\lim_{n \to \infty} p_n(i,j) = 0.$$

Let $\alpha_k(i), k = 1, \ldots, r$ be the probability that the chain starting in state i eventually ends up in recurrent class R_k [in Section 5 we will discuss how to calculate $\alpha_k(i)$]. Once the chain reaches a state in R_k it will settle down to the equilibrium distribution on R_k. From this we see that if $j \in R_k$,

$$\lim_{n \to \infty} p_n(i,j) = \alpha_k(i)\pi^k(j).$$

If $\bar{\phi}$ is an initial probability vector,

$$\lim_{n \to \infty} \bar{\phi} P^n$$

exists but depends on $\bar{\phi}$.

Suppose now that P is irreducible but has period $d > 1$. In this case the state space splits nicely into d sets, $A_1, \ldots A_d$, such that the chain always moves from A_i to A_{i+1} (or A_d to A_1). To illustrate the long-term behavior of P^n, we will consider Example 1, an example with period 2. Let

$$P = \begin{bmatrix} 0 & 1 & 0 & 0 & 0 \\ 1/2 & 0 & 1/2 & 0 & 0 \\ 0 & 1/2 & 0 & 1/2 & 0 \\ 0 & 0 & 1/2 & 0 & 1/2 \\ 0 & 0 & 0 & 1 & 0 \end{bmatrix}.$$

The eigenvalues for P are $1, -1, 0, 1/\sqrt{2}, -1/\sqrt{2}$. The eigenvalue 1 is simple and there is a unique invariant probability $\bar{\pi} = (1/8, 1/4, 1/4, 1/4, 1/8)$. However, when powers of P are taken the eigenvector for -1 becomes important as well as $\bar{\pi}$. We can diagonalize P,

$$D = Q^{-1}PQ,$$

where

$$Q = \begin{bmatrix} 1 & -1/2 & 1/4 & -1 & \sqrt{2}/4 \\ 1 & 1/2 & 0 & -\sqrt{2}/2 & -1/4 \\ 1 & -1/2 & -1/4 & 0 & 0 \\ 1 & 1/2 & 0 & \sqrt{2}/2 & 1/4 \\ 1 & -1/2 & 1/4 & 1 & -\sqrt{2}/4 \end{bmatrix},$$

$$Q^{-1} = \begin{bmatrix} 1/8 & 1/4 & 1/4 & 1/4 & 1/8 \\ -1/4 & 1/2 & -1/2 & 1/2 & -1/4 \\ 1 & 0 & -2 & 0 & 1 \\ -1/4 & -\sqrt{2}/4 & 0 & \sqrt{2}/4 & 1/4 \\ \sqrt{2}/2 & -1 & 0 & 1 & -\sqrt{2}/2 \end{bmatrix},$$

$$D = \begin{bmatrix} 1 & 0 & 0 & 0 & 0 \\ 0 & -1 & 0 & 0 & 0 \\ 0 & 0 & 0 & 0 & 0 \\ 0 & 0 & 0 & 1/\sqrt{2} & 0 \\ 0 & 0 & 0 & 0 & -1/\sqrt{2} \end{bmatrix}.$$

We then see that for P^n, the eigenvectors for the three eigenvalues with absolute value less than 1 become irrelevant and for large n

$$P^n \approx \begin{bmatrix} 1/8 & 1/4 & 1/4 & 1/4 & 1/8 \\ 1/8 & 1/4 & 1/4 & 1/4 & 1/8 \\ 1/8 & 1/4 & 1/4 & 1/4 & 1/8 \\ 1/8 & 1/4 & 1/4 & 1/4 & 1/8 \\ 1/8 & 1/4 & 1/4 & 1/4 & 1/8 \end{bmatrix} +$$

$$(-1)^n \begin{bmatrix} 1/8 & -1/4 & 1/4 & -1/4 & 1/8 \\ -1/8 & 1/4 & -1/4 & 1/4 & -1/8 \\ 1/8 & -1/4 & 1/4 & -1/4 & 1/8 \\ -1/8 & 1/4 & -1/4 & 1/4 & -1/8 \\ 1/8 & -1/4 & 1/4 & -1/4 & 1/8 \end{bmatrix}.$$

Note then that the value of P^n varies depending on whether n is even or odd. In this case the invariant probability at a state i, $\pi(i)$, does not represent the limit of $p_n(j,i)$. However, it does represent the average amount of time that is spent in site i. In fact, one can check that for large n, the average of $p_n(j,i)$ and $p_{n+1}(j,i)$ approaches $\pi(i)$ for each initial state j, i.e.,

$$\pi(i) = \lim_{n\to\infty} \frac{1}{2}(p_n(j,i) + p_{n+1}(j,i)).$$

In general, if P is irreducible with period d, P will have d eigenvalues with absolute value 1, the d complex numbers z with $z^d = 1$. Each is simple; in particular, the eigenvalue 1 is simple and there exists a unique invariant probability $\bar{\pi}$. Given any initial probability distribution $\bar{\phi}$, for large n, $\bar{\phi}P^n$

will cycle through d different distributions, but they will average to $\bar{\pi}$,

$$\lim_{n \to \infty} \frac{1}{d} [\bar{\phi} P^{n+1} + \cdots + \bar{\phi} P^{n+d}] = \bar{\pi}.$$

1.4 Return Times

Let X_n be an irreducible (but perhaps periodic) Markov chain with transition matrix P. Consider the amount of time spent in state j up through time n,

$$Y(j, n) = \sum_{m=0}^{n} I\{X_m = j\}.$$

Here we write I to denote the "indicator function" of an event, i.e., the random variable which equals 1 if the event occurs and 0 otherwise. If $\bar{\pi}$ denotes the invariant probability distribution for P, then it follows from the results of the previous sections that

$$\lim_{n \to \infty} \frac{1}{n+1} E(Y(j, n) \mid X_0 = i) = \lim_{n \to \infty} \frac{1}{n+1} \sum_{m=0}^{n} P\{X_m = j \mid X_0 = i\}$$
$$= \pi(j),$$

i.e., $\pi(j)$ represents the fraction of time that the chain spends in state j. In this section we relate $\pi(j)$ to the first return time to the state j.

Fix a state i and assume that $X_0 = i$. Let T be the first time after 0 that the Markov chain is in state i,

$$T = \min\{n \geq 1 : X_n = i\}.$$

Since the chain is irreducible, we know that $T < \infty$ with probability 1. In fact (see Exercise 1.7) it is not too difficult to show that $E(T) < \infty$.

Consider the time until the kth return to the state i. This time is given by a sum of independent random variables, $T_1 + \cdots + T_k$, each with the distribution of T. Here, T_m denotes the time between the $(m-1)$st and mth return. For k large, the law of large numbers gives that

$$\frac{1}{k}(T_1 + \cdots T_k) \approx E(T),$$

i.e., there are about k visits to the state i in $kE(T)$ steps of the chain. But we have already seen that in n steps we expect about $n\pi(i)$ visits to the state i. Hence setting $n = kE(T)$ we get the relation

$$E(T) = [\pi(i)]^{-1}. \tag{1.11}$$

This says that the expected number of steps to return to i, assuming that the chain starts at i, is given by the inverse of the invariant probability. The above argument is, of course, not completely rigorous, but it does not take too much work to supply the details to prove that (1.11) always holds. See Exercise 1.14 for another derivation of (1.11).

Example 1. Consider the two-state Markov chain with $S = \{0, 1\}$ and

$$P = \begin{bmatrix} 1-p & p \\ q & 1-q \end{bmatrix}, \quad 0 < p, q < 1.$$

Assume the chain starts in state 0 and let T be the return time to 0. In Section 2, we showed that $\bar{\pi} = (q/(p+q), p/(p+q))$ and hence

$$E(T) = [\pi(0)]^{-1} = (p+q)/q. \tag{1.12}$$

In this example we can write down the distribution for T explicitly and verify (1.12). For $n > 1$,

$$P\{T \geq n\} = P\{X_1 = 1, \ldots, X_{n-1} = 1 \mid X_0 = 0\} = p(1-q)^{n-2}.$$

If Y is any random variable taking values in the nonnegative integers,

$$E(Y) = \sum_{n=1}^{\infty} nP\{Y = n\} = \sum_{n=1}^{\infty} \sum_{k=1}^{n} P\{Y = n\}$$

$$= \sum_{k=1}^{\infty} \sum_{n=k}^{\infty} P\{Y = n\} = \sum_{k=1}^{\infty} P\{Y \geq k\}. \tag{1.13}$$

Therefore,

$$E(T) = \sum_{n=1}^{\infty} nP\{T = n\} = \sum_{n=1}^{\infty} P\{T \geq n\}$$

$$= 1 + \sum_{n=2}^{\infty} p(1-q)^{n-2} = (p+q)/q.$$

It should be pointed out that (1.11) only gives the expected value of the random variable T and says nothing else about its distribution. In general, one can say very little else about the distribution of T given only the invariant probability $\bar{\pi}$. For example, consider the two-state example above with $p = q$ so that $E(T) = 2$. If p is close to 1, then $T = 2$ most of the time and $Var(T)$ is small. If p is close to 0, then $T = 1$ most of the time, but occasionally T takes on a very high value. In this case, $Var(T)$ is large.

In the next section, we discuss how to compute the expected number of steps from i to j when $i \neq j$.

1.5 Transient States

Let P be the transition matrix for a Markov chain X_n. Recall that a state i is called transient if with probability 1 the chain visits i only a finite number of times. Suppose P has some transient states and let Q be the submatrix of P which includes only the rows and columns for the transient states. Hence (after rearranging the order of the states) we can write

$$P = \left[\begin{array}{c|c} \tilde{P} & 0 \\ \hline S & Q \end{array}\right], \quad P^n = \left[\begin{array}{c|c} \tilde{P}^n & 0 \\ \hline S_n & Q^n \end{array}\right].$$

As an example, we consider the random walk with absorbing boundaries (Example 2, Section 1.3). We order the state space $S = \{0, 4, 1, 2, 3\}$ so that we can write

$$P = \left[\begin{array}{cc|ccc} 1 & 0 & 0 & 0 & 0 \\ 0 & 1 & 0 & 0 & 0 \\ \hline 1/2 & 0 & 0 & 1/2 & 0 \\ 0 & 0 & 1/2 & 0 & 1/2 \\ 0 & 1/2 & 0 & 1/2 & 0 \end{array}\right]. \qquad (1.14)$$

The matrix Q is a *substochastic matrix*, i.e., a matrix with nonnegative entries whose row sums are less than or equal to 1. Since the states represented by Q are transient, $Q^n \to 0$. This implies that all of the eigenvalues of Q have absolute values strictly less than 1. Hence, $I - Q$ is an invertible matrix and there is no problem in defining the matrix

$$M = (I - Q)^{-1}.$$

Let i be a transient state and consider Y_i, the total number of visits to i,

$$Y_i = \sum_{n=0}^{\infty} I\{X_n = i\}.$$

Since i is transient, $Y_i < \infty$ with probability 1. Suppose $X_0 = j$, where j is another transient state. Then,

$$\begin{aligned} E(Y_i \mid X_0 = j) &= E[\sum_{n=0}^{\infty} I\{X_n = i\} \mid X_0 = j] \\ &= \sum_{n=0}^{\infty} P\{X_n = i \mid X_0 = j\} \\ &= \sum_{n=0}^{\infty} p_n(j, i). \end{aligned}$$

In other words, $E(Y_i \mid X_0 = j)$ is the (j, i) entry of the matrix $I + P + P^2 + \cdots$ which is the same as the (j, i) entry of the matrix $I + Q + Q^2 + \cdots$. However, a simple calculation shows that

$$(I + Q + Q^2 + \cdots)(I - Q) = I,$$

or

$$I + Q + Q^2 + \cdots = (I - Q)^{-1} = M.$$

We have just shown that the expected number of visits to i starting at j is given by M_{ji}, the (j,i) entry of M. If we want to compute the expected number of steps until the chain enters a recurrent class, assuming $X_0 = j$, we need only sum M_{ji} over all transient states i.

In the particular example (1.14),

$$Q = \begin{bmatrix} 0 & 1/2 & 0 \\ 1/2 & 0 & 1/2 \\ 0 & 1/2 & 0 \end{bmatrix}, \quad M = (I - Q)^{-1} = \begin{bmatrix} 3/2 & 1 & 1/2 \\ 1 & 2 & 1 \\ 1/2 & 1 & 3/2 \end{bmatrix}.$$

So, starting, say, in state 1, the expected number of visits to state 3 before absorption is $1/2$, and the expected total number of steps until absorption is $3/2 + 1 + 1/2 = 3$.

We can also use this technique to determine the expected number of steps that an irreducible Markov chain takes to go from one state j to another state i. We first write the transition matritx P for the chain with i being the first site:

$$P = \left[\begin{array}{c|c} p(i,i) & R \\ \hline S & Q \end{array} \right].$$

We then change i to an absorbing site, and hence have the new matrix

$$\tilde{P} = \left[\begin{array}{c|c} 1 & 0 \\ \hline S & Q \end{array} \right].$$

Let T_i be the number of steps needed to reach state i. In other words, T_i is the smallest time n such that $X_n = i$. For any other state k let $T_{i,k}$ be the number of visits to k before reaching i (if we start at state k, we include this as one visit to k). Then,

$$\begin{aligned} E(T_i \mid X_0 = j) &= E\left(\sum_{k \neq i} T_{i,k} \mid X_0 = j \right) \\ &= \sum_{k \neq i} M_{jk}. \end{aligned}$$

In other words, $M\bar{1}$ gives a vector whose jth component is the number of steps starting at j until reaching i.

Example 1. Suppose P is the matrix for random walk with reflecting boundary,

$$P = \begin{bmatrix} 0 & 1 & 0 & 0 & 0 \\ 1/2 & 0 & 1/2 & 0 & 0 \\ 0 & 1/2 & 0 & 1/2 & 0 \\ 0 & 0 & 1/2 & 0 & 1/2 \\ 0 & 0 & 0 & 1 & 0 \end{bmatrix}.$$

If we let $i = 0$, then

$$Q = \begin{bmatrix} 0 & 1/2 & 0 & 0 \\ 1/2 & 0 & 1/2 & 0 \\ 0 & 1/2 & 0 & 1/2 \\ 0 & 0 & 1 & 0 \end{bmatrix}, \quad M = (I - Q)^{-1} = \begin{bmatrix} 2 & 2 & 2 & 1 \\ 2 & 4 & 4 & 2 \\ 2 & 4 & 6 & 3 \\ 2 & 4 & 6 & 4 \end{bmatrix},$$

$$M\bar{1} = (7, 12, 15, 16).$$

Hence, the expected number of steps to get from 4 to 0 is 16.

We now suppose that there are at least two different recurrent classes and ask the question: starting at a given transient state j, what is the probability that the Markov chain eventually ends up in a particular recurrent class? In order to answer this question, we can assume that the recurrent classes consist of single points $r_1, \ldots, r_k$ with $p(r_i, r_i) = 1$. If we order the states so that the recurrent states $r_1, \ldots, r_k$ precede the transient states $t_1, \ldots, t_s$, then

$$P = \left[\begin{array}{c|c} I & 0 \\ \hline S & Q \end{array} \right].$$

For $i = 1, \ldots, s$, $j = 1, \ldots, k$, let $\alpha(t_i, r_j)$ be the probability that the chain starting at t_i eventually ends up in recurrent state r_j. We set $\alpha(r_i, r_i) = 1$ and $\alpha(r_i, r_j) = 0$ if $i \neq j$. For any transient state t_i,

$$\begin{aligned} \alpha(t_i, r_j) &= P\{X_n = r_j \text{ eventually } \mid X_0 = t_i\} \\ &= \sum_{x \in S} P\{X_1 = x \mid X_0 = t_i\} P\{X_n = r_j \text{ eventually } \mid X_1 = x\} \\ &= \sum_{x \in S} p(t_i, x) \alpha(x, r_j). \end{aligned}$$

If A is the $s \times k$ matrix with entries $\alpha(t_i, r_j)$, then the above can be written in matrix form

$$A = S + QA,$$

or

$$A = (I - Q)^{-1} S = MS.$$

Example 2. Consider a random walk with absorbing barrier on $\{0, \ldots, 4\}$. If we order the states $\{0, 4, 1, 2, 3\}$ so that the recurrent states precede the transient states then

$$P = \left[\begin{array}{cc|ccc} 1 & 0 & 0 & 0 & 0 \\ 0 & 1 & 0 & 0 & 0 \\ \hline 1/2 & 0 & 0 & 1/2 & 0 \\ 0 & 0 & 1/2 & 0 & 1/2 \\ 0 & 1/2 & 0 & 1/2 & 0 \end{array} \right],$$

$$S = \left[\begin{array}{cc} 1/2 & 0 \\ 0 & 0 \\ 0 & 1/2 \end{array} \right], \quad M = \left[\begin{array}{ccc} 3/2 & 1 & 1/2 \\ 1 & 2 & 1 \\ 1/2 & 1 & 3/2 \end{array} \right], \quad MS = \left[\begin{array}{cc} 3/4 & 1/4 \\ 1/2 & 1/2 \\ 1/4 & 3/4 \end{array} \right].$$

Hence, starting at state 1 the probability that the the walk is eventually absorbed at state 0 is 3/4.

Example 3. Gambler's Ruin. Here we consider the random walk with absorbing boundary on $\{0, \ldots, N\}$. Let $\alpha(j) = \alpha(j, N)$ be the probability that the walker starting at state j eventually ends up absorbed in state N. Clearly, $\alpha(0) = 0, \alpha(N) = 1$. For $0 < j < N$, we can consider one step as above and note that

$$\alpha(j) = (1 - p)\alpha(j - 1) + p\alpha(j + 1). \tag{1.15}$$

This gives us $N - 1$ linear equations in $N - 1$ unknowns, $\alpha(1), \cdots, \alpha(N - 1)$. To find the solution, we need to know how to solve linear difference equations. By (0.5) and (0.6), the general solution of (1.15) is

$$\alpha(j) = c_1 + c_2 \left(\frac{1 - p}{p} \right)^j, \quad p \neq 1/2,$$

$$\alpha(j) = c_1 + c_2 j, \quad p = 1/2.$$

The boundary conditions $\alpha(0) = 0, \alpha(N) = 1$ allow us to determine the constants c_1, c_2, so we get

$$\alpha(j) = \frac{1 - [(1 - p)/p]^j}{1 - [(1 - p)/p]^N}, \quad p \neq 1/2,$$

$$\alpha(j) = \frac{j}{N}, \quad p = 1/2. \tag{1.16}$$

Note that if $p \leq 1/2$, then for any fixed j,

$$\lim_{N \to \infty} \alpha(j) = 0.$$

This says that if a gambler with fixed resources j plays a fair (or unfair) game in which the gambler wins or loses one unit with each play, then the chance that a gambler will beat a house with very large resources N is very small. However, if $p > 1/2$,

$$\lim_{N \to \infty} \alpha(j) = 1 - \left(\frac{1 - p}{p} \right)^j > 0.$$

This says that there is a positive chance that the gambler playing a game in the gambler's favor will never lose all the resources, and in fact will be able to play forever.

1.6 Examples

Simple Random Walk on a Graph (Example 5, Section 1.1). Assume the graph is connected so that the walk is irreducible. Let e denote the total number of edges in the graph and $d(v)$ the number of edges that have v as one of their endpoints. Since each edge has two endpoints, the sum of $d(v)$ over the vertices in the graph is $2e$. It is easy to check that

$$\pi(v) = d(v)/2e,$$

is the invariant probability measure for this chain.

Urn Model. Suppose there is an urn with N balls. Each ball is colored either red or green. In each time period, one ball is chosen at random from the urn and is replaced with a ball of the other color. Let X_n denote the number of red balls after n picks. Then X_n is an irreducible Markov chain with state space $\{0, \ldots, N\}$. The transition matrix is given by

$$p(j, j+1) = \frac{N-j}{N}, \quad p(j, j-1) = \frac{j}{N}, \quad j = 0, 1, \ldots, N.$$

One might guess that this chain would tend to keep the number of red balls and green balls about the same. In fact, the invariant probability is given by the binomial distribution

$$\pi(j) = \binom{N}{j} 2^{-N}.$$

While it takes some insight to guess that this is correct, it is straightforward to show that in fact this is an invariant probability,

$$
\begin{aligned}
(\bar{\pi} P)(j) &= \sum_{k=0}^{N} \pi(k) p(k, j) \\
&= \pi(j-1) p(j-1, j) + \pi(j+1) p(j+1, j) \\
&= 2^{-N} \binom{N}{j-1} \frac{N - (j-1)}{N} + 2^{-N} \binom{N}{j+1} \frac{j+1}{N} \\
&= 2^{-N} \binom{N}{j} = \pi(j).
\end{aligned}
$$

Hence the probability distribution in equilibrium for the number of red balls is the same as the distribution for the number of heads in N flips of a coin. Recall by the central limit theorem, the number of heads is $N/2$ with a random fluctuation which is of order $\sqrt{N}$.

Cell Genetics. Consider the following Markov chain which models reproduction of cells. Suppose each cell contains N particles each of either one of two types, I or II. Let j be the number of particles of type I. In reproduction, we assume that the cell duplicates itself and then splits, randomly distributing the particles. After duplication, the cell has $2j$ particles oftType I and $2(N - j)$ particles of type II. It then selects N of these $2N$ particles for the next cell. By using the hypergeometric distribution we see that this gives rise to transition probabilities

$$p(j,k) = \frac{\binom{2j}{k}\binom{2(N-j)}{N-k}}{\binom{2N}{N}}.$$

This Markov chain has two absorbing states, 0 and N. Eventually all cells will have only particles of type I or of type II.

Suppose we start with a large number of cells each with j particles of type I. After a long time the population will be full of cells all with one type of particle. What fraction of these will be all type I? Since the fraction of type I particles does not change in this procedure we would expect that the fraction would be j/N. In other words, if we let $\alpha(j)$ be the probability that the Markov chain starting in state j is eventually absorbed in state N, then we expect that

$$\alpha(j) = \frac{j}{N}.$$

For $1 \leq j \leq N - 1$ we can, in fact, verify that this choice of $\alpha(j)$ satisfies

$$\alpha(j) = \sum_{k=0}^{N} p(j,k)\alpha(k),$$

and hence gives the absorption probabilities.

Card Shuffling. Consider a deck of cards numbered $1, \ldots, n$. At each time we will shuffle the cards by drawing a card at random and placing it at the top of the deck. This can be thought of as a Markov chain on S_n, the set of permutations of n elements. If λ denotes any permutation (one-to-one correspondence of $\{1, \ldots, n\}$ with itself), and ν_j denotes the permutation corresponding to moving the jth card to the top of the deck, then the transition probabilities for this chain are given by

$$p(\lambda, \nu_j\lambda) = \frac{1}{n}, \quad j = 1, \ldots, n.$$

This chain is irreducible and aperiodic. It is easy to verify that the unique invariant probability is the uniform measure on S_n, the measure that assigns probability $1/n!$ to each permutation. Therefore, if we start with any ordering of the cards, after enough moves of this kind the deck will be well shuffled.

A much harder question which we will not discuss in this book is how many such moves are "enough" so the deck of cards is shuffled. Other questions, such as the expected number of moves from a given permutation to another given permutation, theoretically can be answered by the methods described in this chapter yet cannot be answered from a practical perspective. The reason in that the transition matrix is $n! \times n!$ which (except for small n) is too large to do the necessary matrix operations.

1.7 Exercises

1.1 The Smiths receive the paper every morning and place it on a pile after reading it. Each afternoon, with probability $1/3$, someone takes all the papers in the pile and puts them in the recycling bin. Also, if ever there are at least five papers in the pile, Mr. Smith (with probability 1) takes the papers to the bin. Consider the number of papers in the pile in the evening. Is it reasonable to model this by a Markov chain? If so, what are the state space and transition matrix?

1.2 Consider a Markov chain with state space $\{0, 1\}$ and transition matrix

$$P = \begin{bmatrix} 1/3 & 2/3 \\ 3/4 & 1/4 \end{bmatrix}.$$

Assuming that the chain starts in state 0 at time $n = 0$, what is the probability that it is in state 1 at time $n = 3$?

1.3 Consider a Markov chain with state space $\{1, 2, 3\}$ and transition matrix

$$P = \begin{bmatrix} .4 & .2 & .4 \\ .6 & 0 & .4 \\ .2 & .5 & .3 \end{bmatrix}.$$

What is the probability in the long run that the chain is in state 1? Solve this problem two different ways: 1) by raising the matrix to a high power; and 2) by directly computing the invariant probability vector as a left eigenvector.

1.4 Do the same for the transition matrix

$$P = \begin{bmatrix} .2 & .4 & .4 \\ .1 & .5 & .4 \\ .6 & .3 & .1 \end{bmatrix}.$$

1.5 Consider the Markov chain with state space $S = \{0, \ldots, 5\}$ and transition matrix

$$P = \begin{bmatrix} .5 & .5 & 0 & 0 & 0 & 0 \\ .3 & .7 & 0 & 0 & 0 & 0 \\ 0 & 0 & .1 & 0 & .9 & 0 \\ .25 & .25 & 0 & 0 & .25 & .25 \\ 0 & 0 & .7 & 0 & .3 & 0 \\ 0 & .2 & 0 & .2 & .2 & .4 \end{bmatrix}.$$

What are the communication classes? Which ones are recurrent and which are transient? Suppose the system starts in state 0. What is the probability that it will be in state 0 at some large time? Answer the same question assuming the system starts in state 5.

1.6 Assume that the chain in Exercise 1.3 starts in state 2. What is the expected number of time intervals until the chain is in state 2 again?

1.7 Let X_n be an irreducible Markov chain on the state space $\{1, \dots, N\}$. Show that there exist $C < \infty$ and $\rho < 1$ such that for any states i, j,

$$P\{X_m \neq j,\ m = 0, \dots, n \mid X_0 = i\} \leq C\rho^n.$$

Show that this implies that $E(T) < \infty$, where T is the first time that the Markov chain reaches the state j. (Hint: there exists a $\delta > 0$ such that for all i, the probability of reaching j some time in the first N steps, starting at i, is greater than δ. Why?)

1.8 Consider simple random walk on the graph below. (Recall that simple random walk on a graph is the Markov chain which at each time moves to an adjacent vertex, each adjacent vertex having the same probability.)

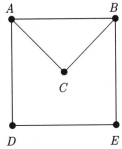

(a) In the long run, about how much time is spent in vertex A?

(b) Suppose a walker starts in vertex A. What is the expected number of steps until the walker returns to A?

(c) Suppose a walker starts in vertex C. What is the expected number of visits to B before the walker reaches A?

(d) Suppose the walker starts in vertex B. What is the probability that the walker reaches A before the walker reaches C?

(e) Again assume the walker starts in C. What is the expected number of steps until the walker reaches A?

1.9 Consider the Markov chain with state space $\{1, 2, 3, 4, 5\}$ and matrix

$$P = \begin{bmatrix} 0 & 1/3 & 2/3 & 0 & 0 \\ 0 & 0 & 0 & 1/4 & 3/4 \\ 0 & 0 & 0 & 1/2 & 1/2 \\ 1 & 0 & 0 & 0 & 0 \\ 1 & 0 & 0 & 0 & 0 \end{bmatrix}.$$

(a) Is the chain irreducible?

(b) What is the period of the chain?

(c) What are $p_{1,000}(2, 1), p_{1,000}(2, 2), p_{1,000}(2, 4)$ (approximately)?

(d) Let T be the first return time to the state 1, starting at state 1. What is the distribution of T and what is $E(T)$? What does this say, without any further calculation, about $\pi(1)$?

(e) Find the invariant probability $\bar\pi$. Use this to find the expected return time to state 2, starting in state 2.

1.10 Let $X_1, X_2, \ldots$ be the successive values from independent rolls of a standard six-sided die. Let $S_n = X_1 + \cdots + X_n$. Let

$$T_1 = \min\{n \geq 1 : S_n \text{ is divisible by } 8\},$$

$$T_2 = \min\{n \geq 1 : S_n - 1 \text{ is divisible by } 8\}.$$

Find $E(T_1)$ and $E(T_2)$. (Hint: consider the remainder of S_n after division by 8 as a Markov chain.)

1.11 Let X_n, Y_n be independent Markov chains with state space $\{0, 1, 2\}$ and transition matrix

$$P = \begin{bmatrix} 1/2 & 1/4 & 1/4 \\ 1/4 & 1/4 & 1/2 \\ 0 & 1/2 & 1/2 \end{bmatrix}.$$

Suppose $X_0 = 0, Y_0 = 2$ and let

$$T = \inf\{n : X_n = Y_n\}.$$

(a) Find $E(T)$.

(b) What is $P\{X_T = 2\}$?

(c) In the long run, what percentage of the time are both chains in the same state?

[Hint: Consider the nine-state Markov chain $Z_n = (X_n, Y_n)$.]

1.12 Consider the Markov chain described in Exercise 1.1.

(a) After a long time, what would be the expected number of papers in the pile?

(b) Assume the pile starts with 0 papers. What is the expected time until the pile will again have 0 papers?

1.13 Let X_n be a Markov chain on state space $\{1, 2, 3, 4, 5\}$ with transition matrix

$$P = \begin{bmatrix} 0 & 1/2 & 1/2 & 0 & 0 \\ 0 & 0 & 0 & 1/5 & 4/5 \\ 0 & 0 & 0 & 2/5 & 3/5 \\ 1 & 0 & 0 & 0 & 0 \\ 1/2 & 0 & 0 & 0 & 1/2 \end{bmatrix}.$$

(a) Is this chain irreducible? Is it aperiodic?

(b) Find the stationary probability vector.

(c) Suppose the chain starts in state 1. What is the expected number of steps until it is in state 1 again?

(d) Again, suppose $X_0 = 1$. What is the expected number of steps until the chain is in state 4?

(e) Again, suppose $X_0 = 1$. What is the probability that the chain will enter state 5 before it enters state 3?

1.14 Let X_n be an irreducible Markov chain with state space S starting at state i with transition matrix P. Let

$$T = \min\{n > 0 : X_n = i\}$$

be the first time that the chain returns to state i. For each state j let $r(j)$ be the expected number of visits to j before returning to i,

$$r(j) = E\Big[\sum_{n=0}^{T-1} I\{X_n = y\}\Big].$$

Note that $r(i) = 1$.

(a) Let $\bar{r}$ be the vector whose jth component is $r(j)$. Show that $\bar{r}P = \bar{r}$.

(b) Show that

$$E(T) = \sum_{j \in S} r(j).$$

(c) Conclude that $E(T) = \pi(i)^{-1}$, where $\bar{\pi}$ denotes the invariant probability.

1.15 In this exercise we outline a proof of the Perron–Frobenius Theorem about matrices with positive entries. Let $A = (a_{ij})$ be an $N \times N$ matrix with $a_{ij} > 0$ for all i, j. For vectors $\bar{u} = (u^1, \ldots, u^N)$ and $\bar{v} = (v^1, \ldots, v^N)$ we write $\bar{u} \geq \bar{v}$ if $u^i \geq v^i$ for each i and $\bar{u} > \bar{v}$ if $u^i > v^i$ for each i. We write $\bar{0} = (0, \ldots, 0)$.

(a) Show that if $\bar{v} \geq \bar{0}$ and $\bar{v} \neq \bar{0}$, then $A\bar{v} > \bar{0}$.

For any vector $\bar{v} \geq \bar{0}$, let $g(\bar{v})$ be the largest λ such that

$$A\bar{v} \geq \lambda\bar{v}.$$

(b) Show that $g(\bar{v}) > 0$ for any nonzero $\bar{v} \geq \bar{0}$ and if $c > 0$ then $g(c\bar{v}) = g(\bar{v})$.

Let
$$\alpha = \sup g(\bar{v}),$$
where the supremum is over all nonzero $\bar{v} \geq 0$. By (b) we can consider the supremum over all v with
$$\|v\| = \sqrt{(v^1)^2 + \cdots + (v^N)^2} = 1.$$
By continuity of the function g on $\{\|v\| = 1\}$ it can be shown that there exists at least one vector $\bar{v} \geq 0$ with $g(\bar{v}) = \alpha$.

(c) Show that for any $\bar{v}$ with $g(\bar{v}) = \alpha$,
$$A\bar{v} = \alpha\bar{v},$$
i.e., $\bar{v}$ is an eigenvector with eigenvalue α. [Hint: we know by definition that $A\bar{v} \geq \alpha\bar{v}$. Assume that they are not equal and consider
$$A[A\bar{v} - \alpha\bar{v}],$$
using (a).]

(d) Show that there is a unique $\bar{v} \geq \bar{0}$ with $g(\bar{v}) = \alpha$ and $\sum_{i=1}^{N} v^i = 1$. [Hint: assume there were two such vectors, $\bar{v}_1, \bar{v}_2$, and consider $g(\bar{v}_1 - \bar{v}_2)$ and $g(|\bar{v}_1 - \bar{v}_2|)$ where
$$|\bar{v}| = (|v^1|, \ldots, |v^n|). \ ]$$

(e) Show that all the components of the $\bar{v}$ in (c) are strictly positive. [Hint: if $A\bar{v} \geq \lambda\bar{v}$ then $A(A\bar{v}) \geq \lambda A\bar{v}$.]

(f) Show that if λ is any other eigenvalue of A, then $|\lambda| < \alpha$. (Hint: assume $A\bar{u} = \lambda\bar{u}$ and consider $A|\bar{u}|$.)

(g) Show that if B is any $(N-1) \times (N-1)$ submatrix of A, then all the eigenvalues of B have absolute value strictly less than α. [Hint: since B is a matrix with positive entries, (a)–(f) apply to B.]

(h) Consider
$$f(\lambda) = \det(A - \lambda I).$$
Show that
$$f'(\lambda) = -\sum_{i=1}^{N} \det(B_i - \lambda I),$$
where B_i denotes the submatrix of A obtained by deleting the ith row and ith column.

(i) Use (g) and (h) to conclude that
$$f'(\alpha) > 0,$$
and hence that α is a simple eigenvalue for A.

(j) Explain why every stochastic matrix with strictly positive entries has a unique invariant probability with all positive components. (Apply the above results to the transpose of the stochastic matrix.)

1.16 An elementary theorem in number theory states that if two integers m and n are relatively prime (i.e., greatest common divisor equal to 1), then there exist integers x and y (positive or negative) such that

$$mx + ny = 1.$$

Using this theorem show the following:

(a) If m and n are relatively prime then the set

$$\{xm + ny : x, y \text{ positive integers }\}$$

contains all but a finite number of the positive integers.

(b) Let J be a set of nonnegative integers whose greatest common divisor is d. Suppose also that J is closed under addition, $m, n \in J \Rightarrow m + n \in J$. Then J contains all but a finite number of integers in the set $\{0, d, 2d, \ldots\}$.

CHAPTER 2

Countable Markov Chains

2.1 Introduction

In this chapter, we consider (time-homogeneous) Markov chains with a countably infinite state space. A set is countably infinite if it can be put into one-to-one correspondence with the set of nonnegative integers $\{0, 1, 2, \ldots\}$. Examples of such sets are: Z, the set of all integers; $2Z$, the set of even integers; and Z^2, the set of lattice points in the plane,

$$Z^2 = \{(i, j) : i, j \text{ integers }\}.$$

(The reader may wish to consider how Z^2 and $\{0, 1, 2, \ldots\}$ can be put into one-to-one correspondence.) Not all infinite sets are countably infinite; for example, the set of real numbers cannot be put into one-to-one correspondence with the positive integers.

We will again let X_n denote a Markov chain. Some of that which was described for finite-state Markov chains holds equally well in the infinite case; however, some things become a bit trickier. We again can speak of the transition matrix, but in this case it becomes an infinite matrix. We will choose not to use the matrix notation here, but simply use the transition probabilities

$$p(x, y) = P\{X_1 = y \mid X_0 = x\}, \quad x, y \in S.$$

The transition probabilities are nonnegative and the "rows" add up to 1, i.e., for each $x \in S$,

$$\sum_{y \in S} p(x, y) = 1.$$

We have chosen to use x, y, z for elements of the state space S. We also define the n-step transition probabilities

$$p_n(x, y) = P\{X_n = y \mid X_0 = x\}.$$

If $0 < m, n < \infty$,

$$
\begin{aligned}
p_{m+n}(x, y) &= P\{X_{m+n} = y \mid X_0 = x\} \\
&= \sum_{z \in S} P\{X_{m+n} = y, X_m = z \mid X_0 = x\}
\end{aligned}
$$

$$= \sum_{z \in S} p_m(x,z) p_n(z,y).$$

This equation is sometimes called the Chapman–Kolmogorov equation.

Example 1. Random Walk with Partially Reflecting Boundary at 0. Let $0 < p < 1$ and $S = \{0,1,2,\ldots\}$. The transition probabilities are given by

$$p(x, x-1) = 1-p, \quad p(x, x+1) = p, \quad x > 0,$$

and

$$p(0,0) = 1-p, \quad p(0,1) = p.$$

Example 2. Simple Random Walk on the Integer Lattice. Let Z^d be the d-dimensional integer lattice, i.e.,

$$Z^d = \{(z_1, \ldots, z_d) : z_i \text{ integers }\}.$$

Note that each element x of Z^d has $2d$ points in Z^d which are distance 1 from x. Simple random walk on Z^d is the process X_n taking values in Z^d which at each time moves to one of the $2d$ nearest neighbors of its current position, choosing equally among all the nearest neighbors. More precisely, it is the Markov chain with state space $S = Z^d$ and

$$p(x,y) = \begin{cases} 1/2d, & \text{if } |x - y| = 1, \\ 0, & \text{otherwise.} \end{cases}$$

Example 3. Queueing Model. Let X_n be the number of customers waiting in line for some service. We think of the first person in line as being serviced while all others are waiting their turn. During each time interval there is a probability p that a new customer arrives. With probability q, the service for the first customer is completed and that customer leaves the queue. We put no limit on the number of customers waiting in line. This is a Markov chain with state space $\{0,1,2,\ldots\}$ and transition probabilities (see Example 2, Section 1.1):

$$p(x, x-1) = q(1-p), \quad p(x,x) = qp + (1-q)(1-p),$$

$$p(x, x+1) = p(1-q), \quad x > 0;$$

$$p(0,0) = 1-p, \quad p(0,1) = p.$$

As in the case of finite Markov chains, our goal will be to understand the behavior for large time. Some of the ideas for finite chains apply equally well to the infinite case. For example, the notion of communication classes applies equally well here. Again, we call a Markov chain irreducible if all the states communicate. All the examples discussed in this chapter are irreducible except for a couple of cases where all the states but one communicate and that one state x is absorbing, $p(x,x) = 1$. We can also talk of the period of an irreducible chain; Examples 1 and 3 above are aperiodic, whereas Example 2 has period 2. It will not always be the case that an irreducible, aperiodic Markov chain with infinite state space converges to an equilibrium distribution.

2.2 Recurrence and Transience

Suppose X_n is an irreducible Markov chain with countably infinite state space S and transition probabilities $p(x, y)$. We say that X_n is a *recurrent* chain if for each state x,

$$P\{X_n = x \text{ for infinitely many } n\} = 1,$$

i.e, if the chain returns infinitely often to x. If an irreducible chain visits a certain state x infinitely often then it must visit every state infinitely often. (The basic reason is that if y is another state there is a positive probability of reaching y from x. If x is visited infinitely often then we get this chance of reaching y infinitely often. If a certain event has a positive probability of occurring, and we get an infinite number of trials, then the event will occur an infinite number of times.) If the chain is not recurrent, then every state is visited only a finite number of times. In this case, the chain is called *transient*. It is often not easy to determine whether a given Markov chain is recurrent or transient. In this section we give two criteria for determining this.

Fix a site x and assume that $X_0 = x$. Consider the random variable R which gives the total number of visits to the site x, including the initial visit. We can write R as

$$R = \sum_{n=0}^{\infty} I\{X_n = x\},$$

where again we use I to denote the indicator function, which equals 1 if the event occurs and 0 otherwise. If the chain is recurrent then R is identically equal to infinity; if the chain is transient, then $R < \infty$ with probability 1. We can compute the expectation of R (assuming $X_0 = x$),

$$E(R) = E \sum_{n=0}^{\infty} I\{X_n = x\} = \sum_{n=0}^{\infty} P\{X_n = x\} = \sum_{n=0}^{\infty} p_n(x, x).$$

We will now compute $E(R)$ in a different way. Let T be the time of first return to x,

$$T = \min\{n > 0 : X_n = x\}.$$

We say that $T = \infty$ if the chain never returns to x. Suppose $P\{T < \infty\} = 1$. Then with probability 1, the chain always returns and by continuing we see that the probability that the chain returns infinitely often is 1 and the chain is recurrent. Now suppose $P\{T < \infty\} = q < 1$, and let us compute the distribution of R in terms of q. First, $R = 1$ if and only if the chain never returns; hence, $P\{R = 1\} = 1 - q$. If $m > 1$, then $R = m$ if and only if the chain returns $m - 1$ times and then does not return for the mth time. Hence, $P\{R = m\} = q^{m-1}(1 - q)$. Therefore, in the transient case, $q < 1$,

$$E(R) = \sum_{m=1}^{\infty} mP\{R = m\} = \sum_{m=1}^{\infty} mq^{m-1}(1 - q) = \frac{1}{1 - q} < \infty.$$

We have concluded the following:

Fact. An irreducible Markov chain is transient if and only if the expected number of returns to a state is finite, i.e., if and only if

$$\sum_{n=0}^{\infty} p_n(x, x) < \infty.$$

Example. Simple Random Walk in Z^d. We first take $d = 1$, and consider the Markov chain on the integers with transition probabilities

$$p(x, x + 1) = p(x, x - 1) = \frac{1}{2}.$$

We will concentrate on the state $x = 0$ and assume $X_0 = 0$. Since this chain has period 2, $p_n(0, 0) = 0$ for n odd. We will write down an exact expression for $p_{2n}(0, 0)$. Suppose the walker is to be at 0 after $2n$ steps. Then the walker must take exactly n steps to the right and n steps to the left. Any "path" of length $2n$ that takes exactly n steps to the right and n steps to the left is equally likely. Each such path has probability $(1/2)^{2n}$ of occurring since it combines $2n$ events each with probability $1/2$. There are $\binom{2n}{n}$ ways of choosing which n of the $2n$ steps should be to the right, and then the other n are to the left. Therefore,

$$p_{2n}(0, 0) = \binom{2n}{n}(\frac{1}{2})^{2n} = \frac{(2n)!}{n!n!}(\frac{1}{2})^{2n}.$$

It is not so easy to see what this looks like for large values of n. However, we can use Stirling's formula to estimate the factorials. Stirling's formula * states that

$$n! \sim \sqrt{2\pi n}\, n^n e^{-n},$$

where $\sim$ means that the ratio of the two sides approaches 1 as n goes to infinity. If we plug this into the above expressions we get that

$$p_{2n}(0, 0) \sim \frac{1}{\sqrt{\pi n}}.$$

---

* See, for example, Feller, *An Introduction to Probability Theory and Its Applications*, Vol, I, II.9.

In particular, since $\sum n^{-1/2} = \infty$,

$$\sum_{n=0}^{\infty} p_{2n}(0,0) = \infty,$$

and simple random walk in one dimension is recurrent.

We now take $d > 1$ so that the chain is on the d-dimensional integer lattice Z^d and has transition probabilities

$$p(x,y) = 1/2d, \quad |x - y| = 1.$$

Again we start the walk at $0 = (0, \ldots, 0)$. We will try to get an asymptotic expression of $p_{2n}(0,0)$ [again $p_n(0,0) = 0$ for n odd]. The combinatorics are somewhat more complicated in this case, so we will give only a sketch of the derivation. Suppose a walker takes $2n$ steps. Then by the law of large numbers, for large values of n, we expect that $2n/d$ of these steps will be taken in each of the d components. We will need the number of steps in each component to be even if we have any chance of being at 0 in n steps. For large n the probability of this occurring is about $(1/2)^{d-1}$ (whether or not an even number of steps have been taken in each of the first $d - 1$ components are almost independent events; however, we know that if an even number of steps have been taken in the first $d - 1$ components then an even number of steps have been taken in the last component as well since the total number of steps taken is even). In each component, if about $2n/d$ steps have been taken, then by the formula above for one dimension, we would expect that the probability that that component equals 0 is about $(\pi(n/d))^{-1/2}$. Combining this, we get an asymptotic expression

$$p_{2n}(0,0) \sim (\frac{1}{2})^{d-1}(\frac{d}{n\pi})^{d/2}.$$

Recall that $\sum n^{-a} < \infty$ if and only if $a > 1$. Hence,

$$\sum_{n=0}^{\infty} p_{2n}(0,0) \begin{cases} = \infty, & d = 1, 2 \\ < \infty, & d \geq 3. \end{cases}$$

Therefore, simple random walk in Z^d is recurrent if $d = 1$ or 2 and is transient if $d \geq 3$.

We now consider another method for determining recurrence or transience. Suppose X_n is an irreducible Markov chain and consider a fixed state which we will denote z. For each state x, we set

$$\alpha(x) = P\{X_n = z \text{ for some } n \geq 0 \mid X_0 = x\}.$$

Clearly, $\alpha(z) = 1$. If the chain is recurrent, then $\alpha(x) = 1$ for all x. However, if the chain is transient there must be states x with $\alpha(x) < 1$. In fact, although not quite as obviously, if the chain is transient there must be points "farther and farther" away from z with $\alpha(x)$ as small as we like.

If $x \neq z$, then

$$
\begin{aligned}
\alpha(x) &= P\{X_n = z \text{ for some } n \geq 0 \mid X_0 = x\} \\
&= P\{X_n = z \text{ for some } n \geq 1 \mid X_0 = x\} \\
&= \sum_{y \in S} P\{X_1 = y \mid X_0 = x\} P\{X_n = z \text{ for some } n \geq 1 \mid X_1 = y\} \\
&= \sum_{y \in S} p(x, y)\alpha(y).
\end{aligned}
$$

Summarizing, $\alpha(x)$ satisfies the following:

$$0 \leq \alpha(x) \leq 1, \tag{2.1}$$

$$\alpha(z) = 1, \quad \inf\{\alpha(x) : x \in S\} = 0, \tag{2.2}$$

and

$$\alpha(x) = \sum_{y \in S} p(x, y)\alpha(y), \quad x \neq z. \tag{2.3}$$

It turns out that if X_n is transient, then there is a unique solution to (2.1) – (2.3) that must correspond to the appropriate probability. [Actually, (2.1) and (2.2) are redundant, but it will be convenient to write them separately.] Moreover, it can be shown (we prove this in Chapter 5, Section 5.5, Example 5) that if X_n is recurrent there is no solution to (2.1) – (2.3). This then gives another possible method to determine recurrence or transience:

Fact. An irreducible Markov chain in transient if and only if for any z we can find a function $\alpha(x)$ satisfying (2.1) – (2.3).

Example. Consider the first example in the last section, random walk with partially reflecting boundary. Let $z = 0$ and let us try to find a solution to (2.1) – (2.3). The third equation states that

$$\alpha(x) = (1 - p)\alpha(x - 1) + p\alpha(x + 1), \quad x > 0.$$

From (0.5) and (0.6) we see that the only solutions to the above equation are of the form

$$\alpha(x) = c_1 + c_2\left(\frac{1 - p}{p}\right)^x, \quad p \neq 1/2,$$

$$\alpha(x) = c_1 + c_2 x, \quad p = 1/2.$$

The first condition in (2.2) gives $\alpha(0) = 1$; plugging this in gives

$$\alpha(x) = (1 - c_2) + c_2\left(\frac{1 - p}{p}\right)^x, \quad p \neq 1/2 \tag{2.4}$$

$$\alpha(x) = 1 + c_2 x, \quad p = 1/2. \tag{2.5}$$

If we choose $c_2 = 0$, we get $\alpha(x) = 1$ for all x which clearly does not satisfy (2.2). If $p = 1/2$ and $c_2 \neq 0$, then the solution is not bounded and hence cannot satisfy (2.1). Similarly, if $p < 1/2$, the solution to (2.4) will be unbounded for $c_2 \neq 0$. In this case, we can conclude that the chain is recurrent for $p \leq 1/2$. For $p > 1/2$, we can find a solution. The second condition in (2.2) essentially boils down to $\alpha(x) \to 0$ as $x \to \infty$, and we get

$$\alpha(x) = (\frac{1-p}{p})^x.$$

Therefore, for $p > 1/2$, the chain is transient.

2.3 Positive Recurrence and Null Recurrence

Suppose X_n is an irreducible, aperiodic Markov chain on the infinite state space S. In this section we investigate when a limiting probability distribution exists. A limiting probability $\pi(x), x \in S$ is a probability distribution on S such that for each $x, y \in S$,

$$\lim_{n \to \infty} p_n(y, x) = \pi(x).$$

If X_n is transient, then

$$\lim_{n \to \infty} p_n(y, x) = 0, \tag{2.6}$$

for all x, y, so no limiting probability distribution exists. It is possible, however, for (2.6) to hold for a recurrent chain. Consider, for example, simple random walk on the integers described in the last section (this is actually a periodic chain, but a small modification can be made to give an aperiodic example). It is recurrent but $p_{2n}(0,0) \to 0$ as $n \to \infty$. We call a chain *null recurrent* if it is recurrent but

$$\lim_{n \to \infty} p_n(x, y) = 0.$$

Otherwise, a recurrent chain is called *positive recurrent*.

Positive recurrent chains behave very similarly to finite Markov chains. If X_n is an irreducible, aperiodic, positive recurrent Markov chain, then for every x, y, the limit

$$\lim_{n \to \infty} p_n(y, x) = \pi(x) > 0,$$

exists and is independent of the initial state y. The $\pi(x)$ give an invariant probability distribution on S, i.e.,

$$\sum_{y \in S} \pi(y) p(y, x) = \pi(x). \tag{2.7}$$

Moreover, if we consider the return time to a state x,

$$T = \min\{n > 0 \mid X_n = x\},$$

then for a positive recurrent chain,

$$E(T \mid X_n = x) = 1/\pi(x).$$

If X_n is null recurrent, then $T < \infty$ with probability 1, but $E(T) = \infty$. (For a transient chain, $T = \infty$ with positive probability.)

One way to determine whether or not a chain is positive recurrent is to try to find an invariant probability distribution. It can be proved that if an irreducible chain is positive recurrent, then there exists a unique probability distribution satisfying (2.7); moreover, if a chain is not positive recurrent, there is no probability distribution satisfying (2.7). This gives a good criterion: try to find an invariant probability distribution. If it exists, then the chain is positive recurrent; if none exists, then it is either null recurrent or transient.

Example. Consider again the example of random walk with partially reflecting boundary. We will try to find a probability distribution that satisfies (2.7), i.e., a nonnegative function $\pi(x)$ satisfying (2.7) and

$$\sum_{x \in S} \pi(x) = 1. \tag{2.8}$$

In this example, (2.7) gives

$$\pi(x+1)(1-p) + \pi(x-1)p = \pi(x), \quad x > 0, \tag{2.9}$$

$$\pi(1)(1-p) + \pi(0)(1-p) = \pi(0). \tag{2.10}$$

By (0.5) and (0.6), the general solution to (2.9) is

$$\pi(x) = c_1 + c_2(\frac{p}{1-p})^x, \quad p \neq 1/2,$$

$$\pi(x) = c_1 + c_2 x, \quad x = 1/2.$$

Equation (2.10) gives $\pi(0) = [(1-p)/p]\pi(1)$. Plugging this into the above gives

$$\pi(x) = c_2(\frac{p}{1-p})^x, \quad p \neq 1/2,$$

$$\pi(x) = c_1, \quad p = 1/2.$$

Now we impose the condition (2.8): can we choose the constant c_1 or c_2 such that $\sum \pi(x) = 1$? For $p = 1/2$, it clearly cannot be done. Suppose $p \neq 1/2$. Clearly, we would need $c_2 \neq 0$. If $p > 1/2$, $\sum [p/(1-p)]^x = \infty$ and we cannot find such a c_2 (we already knew the chain was transient in this case, so it could not possibly be positive recurrent). However if $p < 1/2$, the sum is finite and we can choose

$$\pi(x) = (\frac{p}{1-p})^x [\sum_{y=0}^{\infty} (\frac{p}{1-p})^y]^{-1} = (\frac{1-2p}{1-p})(\frac{p}{1-p})^x.$$

In this case the chain is positive recurrent and this gives the invariant probability. Summarizing the discussion in the last two sections we see that random walk with partially reflecting boundary is

$$\begin{aligned}
\text{positive recurrent} \quad &\text{if } p < 1/2, \\
\text{null recurrent} \quad &\text{if } p = 1/2, \\
\text{transient} \quad &\text{if } p > 1/2.
\end{aligned}$$

2.4 Branching Process

In this section we study a stochastic model for population growth. Consider a system of individuals. We let X_n denote the number of individuals at time n. At each time interval, the population will change according to the following rule: each individual will produce a random number of offspring; after producing the offspring, the individual dies and leaves the system. We make two assumptions about the reproduction process:

1. Each individual produces offspring with the same probability distribution: there are given nonnegative numbers $p_0, p_1, p_2, \ldots$ summing to 1 such that the probability that an individual produces exactly k offspring is p_k.

2. The individuals reproduce independently.

The number of individuals at stage n, X_n, is then a Markov chain with state space $\{0, 1, 2, \ldots\}$. Note that 0 is an absorbing state; once the population dies out, no individuals can be produced. It is not so easy to write down explicitly the transition probabilities for this chain. Suppose that $X_n = k$. Then k individuals produce offspring for the $(n+1)$st generation. If $Y_1, \ldots, Y_k$ are independent random variables each with distribution

$$P\{Y_i = j\} = p_j,$$

then

$$p(k, j) = P\{X_{n+1} = j \mid X_n = k\} = P\{Y_1 + \cdots + Y_k = p_j\}.$$

The actual distribution of $Y_1 + \cdots + Y_k$ can be expressed in terms of convolutions, but we will not need the exact form here. Let μ denote the mean number of offspring produced by an individual,

$$\mu = \sum_{i=0}^{\infty} i p_i.$$

Then,

$$E(X_{n+1} \mid X_n = k) = E(Y_1 + \cdots + Y_k) = k\mu.$$

It is relatively straightforward to calculate the mean number of individuals, $E(X_n)$,

$$E(X_n) = \sum_{k=0}^{\infty} P\{X_{n-1} = k\} E(X_n \mid X_{n-1} = k)$$

$$= \sum_{k=0}^{\infty} k\mu P\{X_{n-1} = k\}$$

$$= \mu E(X_{n-1}).$$

Or, if we do this n times,

$$E(X_n) = \mu^n E(X_0).$$

Some interesting conclusions can be reached from this expression. If $\mu < 1$, then the mean number of offspring goes to 0 as n gets large. The easy estimate

$$E(X_n) = \sum_{k=0}^{\infty} kP\{X_n = k\} \geq \sum_{k=1}^{\infty} P\{X_n = k\} = P\{X_n \geq 1\}$$

can then be used to deduce that the population eventually dies out,

$$\lim_{n \to \infty} P\{X_n = 0\} = 1.$$

If $\mu = 1$, the expected population size remains constant while for $\mu > 1$, the expected population size grows. It is not so clear in these cases whether or not the population dies out with probability 1. [It is possible for X_n to be 0 with probability very near 1, yet $E(X_n)$ not be small.] Below we investigate how to determine the probability that the population dies out. In order to avoid trivial cases we will assume that

$$p_0 > 0; \quad p_0 + p_1 < 1. \tag{2.11}$$

Let

$$a_n(k) = P\{X_n = 0 \mid X_0 = k\}$$

and let $a(k)$ be the probability that the population eventually dies out assuming that there are k individuals initially,

$$a(k) = \lim_{n \to \infty} a_n(k).$$

If the population has k individuals at a certain time, then the only way for the population to die out is for all k branches to die out. Since the branches act independently,

$$a(k) = [a(1)]^k.$$

It suffices therefore to determine $a(1)$ which we will denote by just a and call the *extinction probability*. Assume now that $X_0 = 1$. If we look at one generation, we get

$$a = P\{\text{population dies out} \mid X_0 = 1\}$$

$$= \sum_{k=0}^{\infty} P\{X_1 = k \mid X_0 = 1\} P\{\text{population dies out} \mid X_1 = k\}$$

$$= \sum_{k=0}^{\infty} p_k a(k) = \sum_{k=0}^{\infty} p_k a^k.$$

The quantity on the right is of sufficient interest to give it a name. If X is a random variable taking values in $\{0, 1, 2, \ldots\}$, the *generating function* of X is the function

$$\phi(s) = \phi_X(s) = E(s^X) = \sum_{k=0}^{\infty} s^k P\{X = k\}.$$

Note that $\phi(s)$ is an increasing function of s for $s \geq 0$ with $\phi(0) = P\{X = 0\}$ and $\phi(1) = 1$. Differentiating, we get

$$\phi'(s) = \sum_{k=1}^{\infty} k s^{k-1} P\{X = k\},$$

$$\phi''(s) = \sum_{k=2}^{\infty} k(k-1) s^{k-2} P\{X = k\}.$$

Hence,

$$\phi'(1) = \sum_{k=1}^{\infty} k P\{X = k\} = E(X), \tag{2.12}$$

and for $s > 0$, if $P\{X \geq 2\} > 0$,

$$\phi''(s) > 0. \tag{2.13}$$

If $X_1, \ldots, X_m$ are independent random variables taking values in the non-negative integers, then

$$\phi_{X_1 + \cdots + X_m}(s) = \phi_{X_1}(s) \cdots \phi_{X_m}(s).$$

The easiest way to see this is to use the expression $\phi_X(s) = E(s^X)$ and the product rule for expectation of independent random variables.

Returning to the branching process we see that the extinction probability a satisfies the equation

$$a = \phi(a).$$

Clearly, $a = 1$ satisfies this equation, but there could well be other solutions. Again, we assume $X_0 = 1$. Then the generating function of the random variable X_0 is a and the generating function of X_1 is $\phi(a)$. Let $\phi^n(a)$ be the generating function of X_n. Then $\phi^n(a)$ can be obtained from $\phi^{n-1}(a)$ by $\phi^n(a) = \phi(\phi^{n-1}(a))$. To see this, we first note

$$\phi^n(a) = \sum_{k=0}^{\infty} P\{X_n = k\} a^k$$

$$= \sum_{k=0}^{\infty} [\sum_{j=0}^{\infty} P\{X_1 = j\} P\{X_n = k \mid X_1 = j\}] a^k$$

$$= \sum_{j=0}^{\infty} p_j \sum_{k=0}^{\infty} P\{X_{n-1} = k \mid X_0 = j\} a^k.$$

Now, if $X_0 = j$, then X_{n-1} is the sum of j independent random variables each with the distribution of X_{n-1} given $X_0 = 1$. Hence the sum over k is the generating function of the sum of j independent random variables each with generating function $\phi^{n-1}(a)$ and hence

$$\sum_{k=0}^{\infty} P\{X_{n-1} = k \mid X_0 = j\} a^k = [\phi^{n-1}(a)]^j,$$

and

$$\phi^n(a) = \sum_{j=0}^{\infty} p_j [\phi^{n-1}(a)]^j = \phi(\phi^{n-1}(a)).$$

We now have a recursive way to find $\phi^n(a)$ and hence to find

$$a_n(1) = P\{X_n = 0 \mid X_0 = 1\} = \phi^n(0).$$

We are now ready to demonstrate the following: the extinction probability a is the *smallest* positive root of the equation $a = \phi(a)$. We have already seen that a must satisfy this equation. Let $\hat{a}$ be the smallest positive root. We will show by induction that for every n, $a_n = P\{X_n = 0\} \leq \hat{a}$ (which implies that $a = \lim a_n \leq \hat{a}$). This is obviously true for $n = 0$ since $a_0 = 0$. Assume that $a_{n-1} \leq \hat{a}$. Then

$$P\{X_n = 0\} = \phi^n(0) = \phi(\phi^{n-1}(0)) = \phi(a_{n-1}) \leq \phi(\hat{a}) = \hat{a}.$$

The inequality follows from the fact that ϕ is an increasing function.

Example 1. Suppose $p_0 = 1/4, p_1 = 1/4, p_2 = 1/2$. Then $\mu = 5/4$ and

$$\phi(a) = (1/4) + (1/4)a + (1/2)a^2.$$

Solving $a = \phi(a)$ gives the solutions $a = 1, 1/2$. The extinction probability is $1/2$.

Example 2. Suppose $p_0 = 1/2, p_1 = 1/4, p_2 = 1/4$. Then $\mu = 3/4$ and

$$\phi(a) = (1/2) + (1/4)a + (1/4)a^2.$$

Solving $a = \phi(a)$ gives the solutions $a = 1, 2$. The extinction probability is 1. (We had already demonstrated this fact since $\mu < 1$.)

Example 3. Suppose $p_0 = 1/4, p_1 = 1/2, p_2 = 1/4$. Then $\mu = 1$ and

$$\phi(a) = (1/4) + (1/2)a + (1/4)a^2.$$

Solving $a = \phi(a)$ gives the solutions $a = 1, 1$. The extinction probability is 1.

We finish by establishing a criterion to determine whether or not $a < 1$. We have already seen that if $\mu < 1$, then $a = 1$. Suppose $\mu = 1$. By (2.12), $\phi'(1) = 1$ and therefore by (2.13), $\phi'(s) < 1$ for $s < 1$. Hence for any $s < 1$,

$$1 - \phi(s) = \int_s^1 \phi'(s)ds < 1 - s,$$

i.e., $\phi(s) > s$. Therefore, if $\mu = 1$, the extinction probability is 1. This is an interesting result—even though the expected population size stays at 1, the probability that the population has died out increases to 1. One corollary of this is that the conditional size of the population conditioned that it has not died out must increase with time. That is to say, if one is told at some large time that the population has *not* died out, then one would expect the population to be large.

Now assume $\mu > 1$. Then $\phi'(1) > 1$ and hence there must be some $s < 1$ with $\phi(s) < s$. But $\phi(0) > 0$. By standard continuity arguments, we see that there must be some $a \in (0, s)$ with $\phi(a) = a$. Since $\phi''(s) > 0$ for $s \in (0, 1)$, the curve is convex and there can be at most one $s \in (0, 1)$ with $\phi(s) = s$. In this case, with positive probability the population lives forever. We summarize these results as a theorem.

Theorem. If $\mu \leq 1$, the extinction probability $a = 1$, i.e., the population eventually dies out. If $\mu > 1$, then the extinction probability $a < 1$ and equals the unique root of the equation

$$t = \phi(t),$$

with $0 < t < 1$.

2.5 Exercises

2.1 Consider the queueing model (Example 3). For which values of p, q is the chain null recurrent, positive recurrent, transient?

For the positive recurrent case give the limiting probability distribution π. What is the average length of the queue in equilibrium?

For the transient case, give $\alpha(x) =$ the probability starting at x of ever reaching state 0.

2.2 Consider the following Markov chain with state space $S = \{0, 1, \ldots\}$. A sequence of positive numbers $p_1, p_2, \ldots$ is given with $\sum_{i=1}^{\infty} p_i = 1$. Whenever the chain reaches state 0 it chooses a new state according to the p_i. Whenever the chain is at a state other than 0 it proceeds deterministically, one step at a time, toward 0. In other words, the chain has transition probability

$$p(x, x - 1) = 1, \quad x > 0,$$

$$p(0, x) = p_x, \quad x > 0.$$

This is a recurrent chain since the chain keeps returning to 0. Under what conditions on the p_x is the chain positive recurrent? In this case, what is the limiting probability distribution π? [Hint: it may be easier to compute $E(T)$ directly where T is the time of first return to 0 starting at 0.]

2.3 Consider the Markov chain with state space $S = \{0, 1, 2, \ldots\}$ and transition probabilities:

$$p(x, x+1) = 2/3; \quad p(x, 0) = 1/3.$$

Show that the chain is positive recurrent and give the limiting probability π.

2.4 Consider the Markov chain with state space $S = \{0, 1, 2, \ldots\}$ and transition probabilities:

$$p(x, x+2) = p, \quad p(x, x-1) = 1 - p, \quad x > 0.$$

$$p(0, 2) = p, \quad p(0, 0) = 1 - p.$$

For which values of p is this a transient chain?

2.5 Let X_n be a Markov chain with state space $S = \{0, 1, 2, \ldots\}$. For each of the following transition probabilities, state if the chain is positive recurrent, null recurrent, or transient. If it is positive recurrent, give the stationary probability distribution:
(a) $p(x, 0) = 1/(x+2), \quad p(x, x+1) = (x+1)/(x+2)$;
(b) $p(x, 0) = (x+1)/(x+2), \quad p(x, x+1) = 1/(x+2)$;
(c) $p(x, 0) = 1/(x^2+2) \quad p(x, x+1) = (x^2+1)/(x^2+2)$.

2.6 Given a branching process with the following offspring distributions, determine the extinction probability a.
(a) $p_0 = .25, p_1 = .4, p_2 = .35$.
(b) $p_0 = .5, p_1 = .1, p_3 = .4$.
(c) $p_0 = .91, p_1 = .05, p_2 = .01, p_3 = .01, p_6 = .01, p_{13} = .01$.
(d) $p_i = (1 - q)q^i$, for some $0 < q < 1$.

2.7 Consider the branching process with offspring distribution as in Exercise 2.6(b) and suppose $X_0 = 1$.
(a) What is the probability that the population is extinct in the second generation $(X_2 = 0)$, given that it did not die out in the first generation $(X_1 > 0)$?
(b) What is the probability that the population is extinct in the third generation, given that it was not extinct in the second generation?

2.8 Consider a branching process with offspring distribution given by p_n. We will make the process into an irreducible Markov chain by asserting that if the population ever dies out, then the next generation will have one new individual [in other words, $p(0, 1) = 1$]. For which values of p_n will this chain be positive recurrent, null recurrent, transient?

2.9 Consider the branching process with $p_0 = 1/3, p_1 = 1/3, p_2 = 1/3$. Find, with the aid of a computer, the probability that the population dies out

—in the first 20 steps
—in the first 200 steps
—in the first 2000 steps.

Do the same with values $p_0 = .35, p_1 = .33, p_2 = .32$.

2.10 Let $X_1, X_2, \ldots$ be independent identically distributed random variables taking values in the integers with mean 0. Let $S_0 = 0$ and

$$S_n = X_1 + \cdots + X_n.$$

(a) Let

$$G_n(x) = E[\sum_{j=0}^{n} I\{S_j = n\}]$$

be the expected number of visits to x in the first n steps. Show that for all n and x, $G_n(0) \geq G_n(x)$. (Hint: consider the first j with $S_j = x$.)

(b) Recall that the law of large numbers implies that for each $\epsilon > 0$,

$$\lim_{n \to \infty} P\{|S_n| \leq n\epsilon\} = 1.$$

Show that this implies that for every $\epsilon > 0$,

$$\lim_{n \to \infty} \frac{1}{n} \sum_{|x| \leq \epsilon n} G_n(x) = 1.$$

(c) Using (a) and (b), show that for each $M > \infty$ there is an n such that $G_n(0) \geq M$.

(d) Conclude that S_n is a recurrent Markov chain.

2.11 Let $p_1, p_0, p_{-1}, \ldots$ be a probability distribution on $\{\ldots, -2, -1, 0, 1\}$ with negative mean

$$\sum_n np_n = \mu < 0.$$

Define a Markov chain X_n on the nonnegative integers with transition probabilities

$$p(n, m) = p_{m-n}, \quad m > 0,$$
$$p(n, 0) = \sum_{m \leq 0} p_{m-n}.$$

In other words, X_n acts like a random walk with increments given by the p_i except that the walk is forbidden to jump below 0. The purpose of this exercise is to show that the chain is positive recurrent.

(a) Let $\pi(n)$ be an invariant probability for the chain. Show that for each $n > 0$,

$$\pi(n) = \sum_{m=n-1}^{\infty} \pi(m) p_{n-m}.$$

(b) Let $q_n = p_{1-n}$. Show there exists an $\alpha \in (0, 1)$ such that

$$\alpha = q_0 + q_1\alpha + q_2\alpha^2 + \cdots.$$

(Hint: q_n is the probability distribution of a random variable with mean greater than 1. The right-hand side is the generating function of the q_n.)

(c) Use the α from (b) to find the invariant probability distribution for the chain.

2.12 Let $p(x, y)$ be the transition probability for a Markov chain on a state space S. Call a function f superharmonic at x for p if

$$\sum_{y \in S} p(x, y)f(y) \le f(x).$$

Fix a state $z \in S$.

(a) Let $\mathcal{A}$ be the set of all functions f with $f(z) = 1$; $0 \le f(y) \le 1$ for all $y \in S$; and that are superharmonic at all $y \ne z$. Let g be defined by

$$g(x) = \inf_{f \in \mathcal{A}} f(x).$$

Show that $g \in \mathcal{A}$.

(b) Show that for all $x \ne z$,

$$\sum_{y \in S} p(x, y)g(y) = g(x).$$

[Hint: suppose $\sum_y p(x, y)g(y) < g(x)$ for some x. Show how you can decrease g a little at x so that the function stays superharmonic.]

(c) Let g be as in (a). Show that if $g(x) < 1$ for some x, then

$$\inf_{x \in S} g(x) = 0.$$

[Hint: let $\epsilon = \inf_x g(x)$ and consider $h(x) = (g(x) - \epsilon)/(1 - \epsilon)$.]

(d) Conclude the following: suppose that an irreducible Markov chain with transition probabilities $p(x, y)$ is given and there is a function f that is superharmonic for p at all $y \ne z$; $f(z) = 1$; $0 \le f(y) \le 1$, $y \in S$; and such that $f(x) < 1$ for some $x \in S$. Then the chain is transient.

CHAPTER 3

Continuous-Time Markov Chains

3.1 Poisson Process

Consider X_t the number of customers arriving at a store by time t. Time is now continuous so t takes values in the nonnegative real numbers. Suppose we make three assumptions about the rate at which customers arrive. Intuitively, they are as follows:

1. The number of customers arriving during one time interval does not affect the number arriving during a different time interval.

2. The "average" rate at which customers arrive remains constant.

3. Customers arrive one at a time.

We now make these assumptions mathematically precise. The first assumption is easy: for $s_1 \le t_1 \le s_2 \le t_2 \le \cdots \le s_n \le t_n$, the random variables $X_{t_1} - X_{s_1}, \ldots X_{t_n} - X_{s_n}$ are independent. For the second two assumptions, we first let λ be the rate at which customers arrive, i.e., on the average we expect λt customers in time t. In a small time interval $[t, t + \Delta t]$, we expect that a new customer arrives with probability about $\lambda \Delta t$. The third assumption states that the probability that more than one customer comes in during a small time interval is significantly smaller than this. Rigorously, this becomes

$$P\{X_{t+\Delta t} = X_t\} = 1 - \lambda \Delta t + o(\Delta t), \tag{3.1}$$

$$P\{X_{t+\Delta t} = X_t + 1\} = \lambda \Delta t + o(\Delta t), \tag{3.2}$$

$$P\{X_{t+\Delta t} \ge X_t + 2\} = o(\Delta t). \tag{3.3}$$

Here $o(\Delta t)$ represents some function that is much smaller than Δt for Δt small, i.e.,

$$\lim_{\Delta t \to 0} \frac{o(\Delta t)}{\Delta t} = 0.$$

A stochastic process X_t with $X_0 = 0$ satisfying these assumptions is called a *Poisson process with rate parameter* λ.

We will now determine the distribution of X_t. We will actually derive the distribution in two different ways. First, consider a large number n and write

$$X_t = \sum_{j=1}^{n} [X_{jt/n} - X_{(j-1)t/n}]. \tag{3.4}$$

We have written X_t as the sum of n independent, identically distributed random variables. If n is large, the probability that any of these random variables is 2 or more is small; in fact,

$$P\{X_{jt/n} - X_{(j-1)t/n} \geq 2 \text{ for some } j \leq n\}$$
$$\leq \sum_{j=1}^{n} P\{X_{jt/n} - X_{(j-1)t/n} \geq 2\}$$
$$= nP\{X_{t/n} \geq 2\}.$$

The last term goes to 0 as $n \to \infty$ by (3.3). Hence we can approximate the sum in (3.4) by a sum of independent random variables which equal 1 with probability $\lambda(t/n)$ and 0 with probability $1 - \lambda(t/n)$. By the formula for the binomial distribution,

$$P\{X_t = k\} \approx \binom{n}{k} (\lambda t/n)^k (1 - (\lambda t/n))^{n-k}.$$

Rigorously, we can then show:

$$P\{X_t = k\} = \lim_{n \to \infty} \binom{n}{k} (\lambda t/n)^k (1 - (\lambda t/n))^{n-k}.$$

To take this limit, note that

$$\lim_{n \to \infty} \binom{n}{k} n^{-k} = \lim_{n \to \infty} \frac{n(n-1)\cdots(n-k+1)}{k!n^k} = \frac{1}{k!},$$

and

$$\lim_{n \to \infty} (1 - (\lambda t/n))^{n-k} = \lim_{n \to \infty} (1 - (\lambda t/n))^n \lim_{n \to \infty} (1 - (\lambda t/n))^{-k} = e^{-\lambda t}.$$

Hence,

$$P\{X_t = k\} = e^{-\lambda t} \frac{(\lambda t)^k}{k!},$$

i.e., X_t has a Poisson distribution with parameter λt.

We now derive this formula in a different way. Let

$$P_k(t) = P\{X_t = k\}.$$

Note that $P_0(0) = 1$ and $P_k(0) = 0$, $k > 0$. Equations (3.1) – (3.3) can be used to give a system of differential equations for $P_k(t)$. The definition of the derivative gives

$$P_k'(t) = \lim_{\Delta t \to 0} \frac{1}{\Delta t} (P\{X_{t+\Delta t} = k\} - P\{X_t = k\}).$$

Note that

$$
\begin{aligned}
P\{X_{t+\Delta t} = k\} &= P\{X_t = k\}P\{X_{t+\Delta t} = k \mid X_t = k\} \\
&\quad + P\{X_t = k - 1\}P\{X_{t+\Delta t} = k \mid X_t = k - 1\} \\
&\quad + P\{X_t \le k - 2\}P\{X_{t+\Delta t} = k \mid X_t \le k - 2\} \\
&= P_k(t)(1 - \lambda \Delta t) + P_{k-1}(t)\lambda \Delta t + o(\Delta t).
\end{aligned}
$$

Therefore,

$$
P_k'(t) = \lambda P_{k-1}(t) - \lambda P_k(t).
$$

We can solve these equations recursively. For $k = 0$, the differential equation

$$
P_0'(t) = -\lambda P_0(t), \quad P_0(0) = 1
$$

has the solution

$$
P_0(t) = e^{-\lambda t}.
$$

To solve for $k > 0$ it is convenient to consider

$$
f_k(t) = e^{\lambda t} P_k(t).
$$

Then $f_0(t) = 1$ and the differential equation becomes

$$
f_k'(t) = \lambda f_{k-1}(t), \quad f_k(0) = 0.
$$

It is then easy to check inductively that the solution is

$$
f_k(t) = \lambda^k t^k / k!,
$$

and hence

$$
P_k(t) = e^{-\lambda t} \frac{(\lambda t)^k}{k!},
$$

which is what we derived previously.

Another way to view the Poisson process is to consider the waiting times between customers. Let $T_n, n = 1, 2, \ldots$ be the time between the arrivals of the $(n - 1)$st and nth customers. Let $Y_n = T_1 + \cdots + T_n$ be the total amount of time until n customers arrive. We can write

$$
Y_n = \inf\{t : X_t = n\},
$$

$$
T_n = Y_n - Y_{n-1}.
$$

The T_i should be independent, identically distributed random variables. One property that the T_i should satisfy is the loss of memory property: if we have waited s time units for a customer and no one has arrived, the chance that a customer will come in the next t time units is exactly the same as if there had been some customers before. Mathematically, this property is written

$$
P\{T_i \ge s + t \mid T_i \ge s\} = P\{T_i \ge t\}.
$$

The only real-valued functions satisfying $f(s + t) = f(s)f(t)$ are of the form $f(t) = e^{-bt}$. Hence the distribution of T_i must be an exponential distribution with parameter b. [A random variable Z has an exponential distribution with rate parameter b if it has density

$$f(z) = be^{-bz}, \quad 0 < z < \infty,$$

or equivalently, if it has distribution function

$$F(z) = P\{Z \le z\} = 1 - e^{-bz}, \quad z \ge 0.$$

An easy calculation gives $E(Z) = 1/b$.] It is easy to see what b should be. For large t values we expect for there to be about λt customers. Hence, $Y_{\lambda t} \approx t$. But $Y_n \approx nE(T_i) = n/b$. Hence $\lambda = b$. This gives a means of constructing a Poisson process: take independent random variables $T_1, T_2, \ldots$, each exponential with rate λ, and define

$$Y_n = T_1 + \cdots + T_n,$$

$$X_t = n, \quad \text{if } Y_n \le t < Y_{n+1}.$$

From this we could then conclude in a third way that the random variables X_t have a Poisson distribution. However, given that we already have the Poisson process, it is easy to compute the distribution of T_i since

$$P\{T_1 > t\} = P\{X_t = 0\} = e^{-\lambda t}.$$

3.2 Finite State Space

In this section we discuss continuous-time Markov chains on a finite state space. We start by discussing some facts about exponential random variables. Suppose $T_1, \ldots, T_n$ are independent random variables, each exponential with rates $b_1, \ldots, b_n$. Intuitively, we can think of n alarm clocks. Consider the first time when any of the alarm clocks goes off; more precisely, consider the random variable

$$T = \min\{T_1, \ldots, T_n\}.$$

Note that

$$
\begin{aligned}
P\{T \ge t\} &= P\{T_1 \ge t, \ldots, T_n \ge t\} \\
&= P\{T_1 \ge t\}P\{T_2 \ge t\} \cdots P\{T_n \ge t\} \\
&= e^{-b_1 t}e^{-b_2 t} \cdots e^{-b_n t} = e^{-(b_1 + \cdots + b_n)t}.
\end{aligned}
$$

In other words, T has an exponential distribution with parameter $b_1 + \cdots + b_n$. Moreover, it is easy to give the probabilities for which of the clocks goes off first,

$$
\begin{aligned}
P\{T_1 = T\} &= \int_0^\infty P\{T_2 > t, \ldots, T_n > t\}\, dP\{T_1 = t\} \\
&= \int_0^\infty e^{-(b_2 + \cdots + b_n)t} b_1 e^{-b_1 t}\, dt
\end{aligned}
$$

$$= \frac{b_1}{b_1 + \cdots + b_n}.$$

In other words, the probability that the ith clock goes off first is the ratio of b_i to $b_1 + \cdots + b_n$. If we are given an infinite sequence of exponential random variables $T_1, T_2, \ldots$, with parameters $b_1, b_2, \ldots$, the same result holds provided that $b_1 + b_2 + \cdots < \infty$.

Suppose now that we have a finite state space S. We will define a process X_t that has the Markov property,

$$P\{X_t = y \mid X_r, 0 \leq r \leq s\} = P\{X_t = y \mid X_s\},$$

and that is time-homogeneous,

$$P\{X_t = y \mid X_s = x\} = P\{X_{t-s} = y \mid X_0 = x\}.$$

For each $x, y \in S, x \neq y$ we assign a nonnegative number $\alpha(x, y)$ that we think of as the rate at which the chain changes from state x to state y. We let $\alpha(x)$ denote the total rate at which the chain is changing from state x, i.e.,

$$\alpha(x) = \sum_{y \neq x} \alpha(x, y).$$

A (time-homogeneous) continuous-time Markov chain with rates α is a stochastic process X_t taking values in S satisfying

$$P\{X_{t+\Delta t} = x \mid X_t = x\} = 1 - \alpha(x)\Delta t + o(\Delta t), \qquad (3.5)$$

$$P\{X_{t+\Delta t} = x \mid X_t = y\} = \alpha(y, x)\Delta t + o(\Delta t), \quad y \neq x. \qquad (3.6)$$

In other words, the probability that the chain in state y jumps to a different state x in a small time interval of length Δt is about $\alpha(y, x)\Delta t$. For the Poisson process, we used the description for small Δt to write differential equations for the probabilities. We do the same in this case. If we let $p_x(t) = P\{X_t = x\}$, then the equations above can be shown to give a system of linear differential equations,

$$p'_x(t) = -\alpha(x)p_x(t) + \sum_{y \neq x} \alpha(y, x)p_y(t).$$

If we impose an initial condition, $p_x(0), x \in S$, then we can solve the system. This system is often written in matrix form. Let A be the matrix whose (x, y) entry equals $\alpha(x, y)$ if $x \neq y$ and equals $-\alpha(x)$ if $x = y$. Then if $\bar{p}(t)$ denotes the vector of probabilities, the system can be written

$$\bar{p}'(t) = \bar{p}(t)A. \qquad (3.7)$$

The matrix A is called the *infinitesimal generator* of the chain. Note that the row sums of A equal 0, the nondiagonal entries of A are nonnegative, and the diagonal entries are nonpositive. From differential equations (see

Chapter 0, Section 0.2), we can give the solution

$$\bar{p}(t) = \bar{p}(0)e^{tA}.$$

We can also write this in terms of transition matrices. Let $p_t(x, y) = P\{X_t = y \mid X_0 = x\}$ and let P_t be the matrix whose (x, y) entry is $p_t(x, y)$. The system of differential equations can be written as a single matrix equation:

$$\frac{d}{dt}P_t = P_t A, \quad P_0 = I. \tag{3.8}$$

The matrix P_t is then given by

$$P_t = e^{tA}.$$

Example 1. Consider a chain with two states—$0, 1$. Assume $\alpha(0, 1) = 1$ and $\alpha(1, 0) = 2$. Then the infinitesimal generator is

$$A = \begin{bmatrix} -1 & 1 \\ 2 & -2 \end{bmatrix}.$$

In order to compute e^{tA}, we diagonalize the matrix. The eigenvalues are $0, -3$. We can write

$$D = Q^{-1}AQ,$$

where

$$D = \begin{bmatrix} 0 & 0 \\ 0 & -3 \end{bmatrix}, Q = \begin{bmatrix} 1 & 1 \\ 1 & -2 \end{bmatrix}, Q^{-1} = \begin{bmatrix} 2/3 & 1/3 \\ 1/3 & -1/3 \end{bmatrix}.$$

We use the diagonalization to compute the exponential e^{tA}.

$$
\begin{aligned}
P_t = e^{tA} &= \sum_{n=0}^{\infty} \frac{(tA)^n}{n!} \\
&= \sum_{n=0}^{\infty} \frac{Q(tD)^n Q^{-1}}{n!} \\
&= Q \begin{bmatrix} 1 & 0 \\ 0 & e^{-3t} \end{bmatrix} Q^{-1} \\
&= \begin{bmatrix} 2/3 & 1/3 \\ 2/3 & 1/3 \end{bmatrix} + e^{-3t} \begin{bmatrix} 1/3 & -1/3 \\ -2/3 & 2/3 \end{bmatrix}.
\end{aligned}
$$

Note that

$$\lim_{t\to\infty} P_t = \begin{bmatrix} \bar{\pi} \\ \bar{\pi} \end{bmatrix},$$

where $\bar{\pi} = (2/3, 1/3)$.

Example 2. Consider a chain with four states—$0, 1, 2, 3$—and infinitesimal generator

$$A = \begin{bmatrix} -1 & 1 & 0 & 0 \\ 1 & -3 & 1 & 1 \\ 0 & 1 & -2 & 1 \\ 0 & 1 & 1 & -2 \end{bmatrix}.$$

The eigenvalues of A are $0, -1, -3, -4$ with right eigenvectors (which are left eigenvectors as well since A is symmetric) $(1, 1, 1, 1)$, $(1, 0, -1/2, -1/2)$, $(0, 0, -1/2, 1/2)$, and $(-1/3, 1, -1/3, 1/3)$. Then,

$$D = Q^{-1}DQ,$$

where

$$D = \begin{bmatrix} 0 & 0 & 0 & 0 \\ 0 & -1 & 0 & 0 \\ 0 & 0 & -3 & 0 \\ 0 & 0 & 0 & -4 \end{bmatrix}, \quad Q = \begin{bmatrix} 1 & 1 & 0 & -1/3 \\ 1 & 0 & 0 & 1 \\ 1 & -1/2 & -1/2 & -1/3 \\ 1 & -1/2 & 1/2 & -1/3 \end{bmatrix},$$

$$Q^{-1} = \begin{bmatrix} 1/4 & 1/4 & 1/4 & 1/4 \\ 2/3 & 0 & -1/3 & -1/3 \\ 0 & 0 & -1 & 1 \\ -1/4 & 3/4 & -1/4 & -1/4 \end{bmatrix}.$$

Therefore,

$$P_t = e^{tA} = Qe^{tD}Q^{-1} =$$

$$\begin{bmatrix} 1/4 & 1/4 & 1/4 & 1/4 \\ 1/4 & 1/4 & 1/4 & 1/4 \\ 1/4 & 1/4 & 1/4 & 1/4 \\ 1/4 & 1/4 & 1/4 & 1/4 \end{bmatrix} + e^{-t}\begin{bmatrix} 2/3 & 0 & -1/3 & -1/3 \\ 0 & 0 & 0 & 0 \\ -1/3 & 0 & 1/6 & 1/6 \\ -1/3 & 0 & 1/6 & 1/6 \end{bmatrix}$$

$$+ e^{-3t}\begin{bmatrix} 0 & 0 & 0 & 0 \\ 0 & 0 & 0 & 0 \\ 0 & 0 & 1/2 & -1/2 \\ 0 & 0 & -1/2 & 1/2 \end{bmatrix} + e^{-4t}\begin{bmatrix} 1/12 & -1/4 & 1/12 & 1/12 \\ -1/4 & 3/4 & -1/4 & -1/4 \\ 1/12 & -1/4 & 1/12 & 1/12 \\ 1/12 & -1/4 & 1/12 & 1/12 \end{bmatrix}.$$

Note that

$$\lim_{t \to \infty} P_t = \begin{bmatrix} 1/4 & 1/4 & 1/4 & 1/4 \\ 1/4 & 1/4 & 1/4 & 1/4 \\ 1/4 & 1/4 & 1/4 & 1/4 \\ 1/4 & 1/4 & 1/4 & 1/4 \end{bmatrix}.$$

We can use exponential waiting times to give an alternative description of the Markov chain. Suppose rates $\alpha(x, y)$ have been given. Suppose $X_0 = x$. Let

$$T = \inf\{t : X_t \neq x\},$$

i.e., T is the time at which the process first changes state. The Markov property can be used to see that T must have the loss of memory property, and hence T must have an exponential distribution. By (3.5),

$$P\{T \leq \Delta t\} = \alpha(x)\Delta t + o(\Delta t).$$

In order for this to be true, T must be exponential with parameter $\alpha(x)$. What state does the chain move to? The infinitesimal characterization (3.6) can be used to check that the probability that the state changes to y is exactly $\alpha(x,y)/\alpha(x)$. By the discussion of exponential distributions above we can think of this in another way. Independent "alarm clocks" are placed at each state y, with each alarm going off at an exponential time with rate $\alpha(x,y)$. The chain stays in state x until the first such clock goes off and then it moves to the state corresponding to that clock.

As in the case for discrete time, we are interested in the large time behavior. As Examples 1 and 2 in this section demonstrate we expect

$$\lim_{t\to\infty} P_t = \Pi_t = \begin{bmatrix} \bar{\pi} \\ \vdots \\ \bar{\pi} \end{bmatrix},$$

where $\bar{\pi}$ represents a limiting probability. The limiting probability should not change with time; hence, by (3.7),

$$\bar{\pi}A = \bar{0}.$$

In this case, $\bar{\pi}$ is an eigenvector of A with eigenvalue 0. The limit theory now parallels that for discrete time. Suppose for ease that the chain is irreducible. [We say that the chain is irreducible if all states communicate, i.e., for each $x, y \in S$, there exist $z_1, \ldots, z_j \in S$ with $\alpha(x, z_1)$, $\alpha(z_1, z_2), \ldots,$ $\alpha(z_{j-1}, z_j), \alpha(z_j, y)$ all strictly positive.] In this case, one can show (see Exercise 3.4) using the results for stochastic matrices that:

1. There is a unique probability vector $\bar{\pi}$ satisfying

$$\bar{\pi}A = \bar{0}.$$

2. All other eigenvalues of A have negative real part.

Then by analyzing the matrix differential equation it is not too difficult to show that

$$\lim_{t\to\infty} P_t = \begin{bmatrix} \bar{\pi} \\ \vdots \\ \bar{\pi} \end{bmatrix}.$$

If the chain is reducible, we must analyze the chain on each communication class. We have not discussed periodicity. This phenomenon does not occur for continuous-time chains; in fact, one can prove (see Exercise 3.6) that for any irreducible continuous-time chain, P_t has strictly positive entries for all $t > 0$.

A number of the methods for analyzing discrete-time chains have analogues for continuous-time chains. Suppose X_t is an irreducible continuous-time chain on finite state space S and suppose z is some fixed state in S. We will compute the mean passage time to z starting at state x, i.e., $b(x) = E(Y \mid X_0 = x)$, where

$$Y = \inf\{t : X_t = z\}.$$

Clearly, $b(z) = 0$. For $x \neq z$, assume $X_0 = x$ and let T be the first time that the chain changes state as above. Then

$$E(Y \mid X_0 = x) = E(T \mid X_0 = x) + \sum_{y \in S} P\{X_T = y \mid X_0 = x\} E(Y \mid X_0 = y).$$

Since T is exponential with parameter $\alpha(x)$ the first term on the right hand side equals $1/\alpha(x)$. Also from the above discussion, $P\{X_t = y \mid X_0 = x\} = \alpha(x,y)/\alpha(x)$. Finally, since $b(z) = 0$, we do not need to include the $y = z$ term in the sum. Therefore, the equation becomes

$$\alpha(x)b(x) = 1 + \sum_{y \neq x, z} \alpha(x,y)b(y).$$

If we let $\tilde{A}$ be the matrix obtained from A by deleting the row and column associated to the state z, we get the matrix equation

$$\bar{0} = \bar{1} + \tilde{A}\bar{b},$$

or

$$\bar{b} = [-\tilde{A}]^{-1}\bar{1}.$$

(The matrix $\tilde{A}$ is a square matrix whose row sums are all nonpositive and at least one of whose row sums is strictly negative. From this one can conclude that all the eigenvalues of $\tilde{A}$ have strictly negative real part, and hence $\tilde{A}$ is invertible.)

Example 3. Consider Example 2 in this section and let us compute the expected time to get from state 0 to state 3. Then $z = 3$,

$$\tilde{A} = \begin{bmatrix} -1 & 1 & 0 \\ 1 & -3 & 1 \\ 0 & 1 & -2 \end{bmatrix},$$

and

$$\bar{b} = [-\tilde{A}]^{-1}\bar{1} = (8/3, 5/3, 4/3).$$

Therefore the expected time to get from state 0 to state 3 is $8/3$.

3.3 Birth-and-Death Processes

In this section we consider a large class of infinite state space, continuous-time Markov chains that are known by the name of *birth-and-death pro-*

cesses. The state space will be $\{0, 1, 2, \ldots\}$, and changes of state will always be from n to $n + 1$ or n to $n - 1$. Intuitively we can view the state of the system as the size of a population that can increase or decrease by 1 by a "birth" or a "death," respectively. To describe the chain, we give birth rates $\lambda_n, n = 0, 1, 2, \ldots$ and death rates $\mu_n, n = 1, 2, 3, \ldots$. If the population is currently n, then new individuals arrive at rate λ_n and individuals leave at rate μ_n (note if the population is 0 there can be no deaths, so $\mu_0 = 0$).

If we let X_t denote the state of the chain at time t, then

$$P\{X_{t+\Delta t} = n \mid X_t = n\} = 1 - (\mu_n + \lambda_n)\Delta t + o(\Delta t),$$

$$P\{X_{t+\Delta t} = n + 1 \mid X_t = n\} = \lambda_n \Delta t + o(\Delta t), \text{ Birth}$$

$$P\{X_{t+\Delta t} = n - 1 \mid X_t = n\} = \mu_n \Delta t + o(\Delta t). \text{ death}$$

As before, we can convert these equations into differential equations for $P_n(t) = P\{X_t = n\}$ and get the system

$$P_n'(t) = \mu_{n+1} P_{n+1}(t) + \lambda_{n-1} P_{n-1}(t) - (\mu_n + \lambda_n) P_n(t). \qquad (3.9)$$

To compute the transition probabilities

$$p_t(m, n) = P\{X_t = n \mid X_0 = m\}$$

we need only solve the system with initial conditions,

$$P_m(0) = 1, \quad P_n(0) = 0, \quad n \neq m.$$

Example 1. The Poisson process with rate parameter λ is a birth-and-death process with $\lambda_n = \lambda$ and $\mu_n = 0$.

Example 2. Markovian Queueing Models. Suppose X_t denotes the number of people on line for some service. We assume that people arrive at a rate λ; more precisely, the arrival rate of customers follows a Poisson process with rate λ. Customers are also serviced at an exponential rate μ. We note three different service rules:

(a) $M/M/1$ queue. In this case there is one server and only the first person in line is being serviced. This gives a birth-and-death process with $\lambda_n = \lambda$ and $\mu_n = \mu$ $(n \geq 1)$. The two Ms in the notation refer to the fact that both the arrival and the service times are exponential and hence the process is Markovian. The 1 denotes the fact that there is one server.

(b) $M/M/k$ queue. In this case there are k servers and anyone in the first k positions in the line can be served. If there are k people being served, and each one is served at rate μ, then the rate at which at people are leaving the system is $k\mu$. This gives a birth-and-death process with $\lambda_n = \lambda$ and

$$\mu_n = \begin{cases} n\mu, & \text{if } n \leq k, \\ k\mu, & \text{if } n \geq k. \end{cases}$$

(c) $M/M/\infty$ queue. In this case there are an infinite number of servers, so everyone in line has a chance of being served. In this case $\lambda_n = \lambda$ and $\mu_n = n\mu$.

Example 3. Population Model. Imagine that the state of the chain represents the number of individuals in a population. Each individual at a certain rate λ produces another individual. Similarly each individual dies at rate μ. If all the individuals act independently this can be modelled by a birth-and-death process with $\lambda_n = n\lambda$ and $\mu_n = n\mu$. Note that 0 is an absorbing state in this model. When $\mu = 0$, this is sometimes called the Yule process.

Example 4. Population Model with Immigration. Assume that individuals die and reproduce with rates μ and λ, respectively, as in the previous model. We also assume that new individuals arrive at a constant rate ν. This gives a birth-and-death process with $\lambda_n = n\lambda + \nu$ and $\mu_n = n\mu$.

Example 5. Fast-Growing Population Model. Imagine that a population grows at a rate proportional to the square of the number of individuals. Then if we assume no deaths, we have a process with $\lambda_n = n^2\lambda$ and $\mu_n = 0$. The population in this case grows very fast, and as seen later it actually reaches an "infinite population" in finite time.

We will look more closely at all of these examples, but first we develop some general theory. We call the birth-and-death chain irreducible if all the states communicate. It is not very difficult to see that this happens if and only if all the λ_n ($n \geq 0$) and all the μ_n ($n \geq 1$) are positive. An irreducible chain is recurrent if one always returns to a state; otherwise, it is called transient. For any birth-and-death process, there is a discrete-time Markov chain on $\{0, 1, 2, \ldots\}$ that follows the continuous-time chain "when it moves." It has transition probabilities

$$p(n, n-1) = \frac{\mu_n}{\mu_n + \lambda_n}, \quad p(n, n+1) = \frac{\lambda_n}{\mu_n + \lambda_n}.$$

One can check that the continuous-time chain is recurrent if and only if the corresponding discrete-time chain is recurrent. Let $a(n)$ be the probability that the chain starting at state n ever reaches state 0. Note that $a(0) = 1$ and the value of $a(n)$ is the same whether one considers the continuous-time or the discrete-time chain. From our discussion of discrete-time chains, we see that $a(n)$ satisfies

$$a(n)(\mu_n + \lambda_n) = a(n-1)\mu_n + a(n+1)\lambda_n, \quad n > 0. \qquad (3.10)$$

If the chain is transient, $a(n) \to 0$ as $n \to \infty$. If the chain is recurrent, no solution of this equation will exist with $a(0) = 1, 0 \leq a(n) \leq 1, a(n) \to 0$ $(n \to \infty)$.

We now give a necessary and sufficient condition for a birth-and-death chain to be transient. We will try to find the function $a(n)$. The equation (3.10) can be rewritten

$$a(n) - a(n+1) = \frac{\mu_n}{\lambda_n}[a(n-1) - a(n)], \quad n \geq 1.$$

If we continue, we get

$$a(n) - a(n+1) = \frac{\mu_1 \cdots \mu_n}{\lambda_1 \cdots \lambda_n}[a(0) - a(1)].$$

Hence,

$$
\begin{aligned}
a(n+1) &= [a(n+1) - a(0)] + a(0) \\
&= \sum_{j=0}^{n}[a(j+1) - a(j)] + 1 \\
&= [a(1) - 1]\sum_{j=0}^{n}\frac{\mu_1 \cdots \mu_j}{\lambda_1 \cdots \lambda_j} + 1,
\end{aligned}
$$

where the $j = 0$ term of the sum equals 1 by convention. We can find a nontrivial solution if the sum converges, i.e., the birth-and-death chain is transient if and only if

$$\sum_{n=1}^{\infty}\frac{\mu_1 \cdots \mu_n}{\lambda_1 \cdots \lambda_n} < \infty. \tag{3.11}$$

As an example, consider the queueing models (Example 2). For the $M/M/1$ queue,

$$\sum_{n=1}^{\infty}\frac{\mu_1 \cdots \mu_n}{\lambda_1 \cdots \lambda_n} = \sum_{n=1}^{\infty}(\frac{\mu}{\lambda})^n,$$

which converges if and only if $\mu < \lambda$. Consider now the $M/M/k$ queue. For any $n \geq k$,

$$\frac{\mu_1 \cdots \mu_n}{\lambda_1 \cdots \lambda_n} = (\frac{k!}{k^k})(\frac{k\mu}{\lambda})^n.$$

Therefore, in this case the sum is finite and the chain is transient if and only if $k\mu < \lambda$. Finally for the $M/M/\infty$ queue,

$$\sum_{n=1}^{\infty}\frac{\mu_1 \cdots \mu_n}{\lambda_1 \cdots \lambda_n} = \sum_{n=1}^{\infty}n!(\frac{\mu}{\lambda})^n = \infty.$$

Hence, for all values of μ and λ the chain is recurrent. These three results can be summarized by saying that the queueing models are transient (and hence the lines grow longer and longer) if and only if the (maximal) service rate is strictly less than the arrival rate.

For recurrent chains, there may or may not be a limiting probability. Again, we call an irreducible chain positive recurrent if there exists a probability distribution $\pi(n)$ such that

$$\lim_{t \to \infty} P\{X_t = n \mid X_0 = m\} = \pi(n).$$

for all states m. Otherwise a recurrent chain is called null recurrent. If the system is in the limiting probability, i.e., if $P_n(t) = \pi(n)$, where $P_n(t)$ is as in (3.9), then $P_n'(t)$ should equal 0. In other words π should satisfy

$$0 = \lambda_{n-1}\pi(n-1) + \mu_{n+1}\pi(n+1) - (\lambda_n + \mu_n)\pi(n). \qquad (3.12)$$

Again, as for the case of discrete-time chains, we can find π by solving these equations. If we can find a probability distribution that satisfies (3.12), then the chain is positive recurrent and that distribution is the unique equilibrium distribution.

We can solve (3.12) directly. First, the equation for $n = 0$ gives

$$\pi(1) = \frac{\lambda_0}{\mu_1}\pi(0).$$

For $n \geq 1$, the equation can be written

$$\mu_{n+1}\pi(n+1) - \lambda_n\pi(n) = \mu_n\pi(n) - \lambda_{n-1}\pi(n-1).$$

If we iterate this equation, we get

$$\mu_{n+1}\pi(n+1) - \lambda_n\pi(n) = \mu_1\pi(1) - \lambda_0\pi(0) = 0.$$

Hence, $\pi(n+1) = (\lambda_n/\mu_{n+1})\pi(n)$, and by iterating we get the solution

$$\pi(n) = \frac{\lambda_0 \cdots \lambda_{n-1}}{\mu_1 \cdots \mu_n}\pi(0).$$

We now impose the condition that π be a probability measure. We can arrange this if and only if $\sum \pi(x) < \infty$. Therefore, a birth-and-death chain is positive recurrent if and only if

$$q = \sum_{n=0}^{\infty} \frac{\lambda_0 \cdots \lambda_{n-1}}{\mu_1 \cdots \mu_n} < \infty$$

(by convention, the $n = 0$ term in this sum is equal to 1). In this case the invariant probability is given by

$$\pi(n) = \frac{\lambda_0 \cdots \lambda_{n-1}}{\mu_1 \cdots \mu_n}q^{-1}. \qquad (3.13)$$

As an example, consider the queueing models again. For the $M/M/1$ queue,

$$\sum_{n=0}^{\infty} \frac{\lambda_0 \cdots \lambda_{n-1}}{\mu_1 \cdots \mu_n} = \sum_{n=0}^{\infty} (\frac{\lambda}{\mu})^n = [1 - (\lambda/\mu)]^{-1},$$

provided $\lambda < \mu$ and is infinite otherwise. Hence this chain is positive recurrent for $\lambda < \mu$ in which case the equilibrium distribution is

$$\pi(n) = (1 - \frac{\lambda}{\mu})(\frac{\lambda}{\mu})^n.$$

Note that the expected length of the queue in equilibrium is

$$\sum_{n=0}^{\infty} n\pi(n) = \sum_{n=0}^{\infty} n(1 - \frac{\lambda}{\mu})(\frac{\lambda}{\mu})^n = (\lambda/\mu)[1 - (\lambda/\mu)]^{-1} = \lambda/(\mu - \lambda).$$

In particular, the expected length gets large as λ approaches μ. In the case of the $M/M/k$ queue, the exact form of π is a little messy, but it is easy to verify that the chain is positive recurrent if and only if $\lambda < k\mu$. Finally for the $M/M/\infty$ queue,

$$\sum_{n=0}^{\infty} \frac{\lambda_0 \cdots \lambda_{n-1}}{\mu_1 \cdots \mu_n} = \sum_{n=0}^{\infty} \frac{1}{n!}(\frac{\lambda}{\mu})^n = e^{\lambda/\mu}.$$

Hence, the chain is positive recurrent for all λ, μ and has equilibrium distribution

$$\pi(n) = e^{-\lambda/\mu} \frac{(\lambda/\mu)^n}{n!},$$

i.e., the equilibrium distribution is a Poisson distribution with parameter λ/μ. The mean queue length in equilibrium is λ/μ.

Conditions under which the population models are positive recurrent, null recurrent, or transient are discussed in Exercises 3.9 and 3.10.

We finish by considering two pure birth processes. A birth-and-death process is a pure birth process if $\mu_n = 0$ for all n. We first consider the Yule process with $\lambda_n = n\lambda$. Let us assume that the population starts with one individual; hence, $P_1(0) = 1, P_n(t) = 0$ $(n > 1)$, where again $P_n(t) = P\{X_t = n\}$. The $P_n(t)$s satisfy the differential equations

$$P_n'(t) = (n - 1)\lambda P_{n-1}(t) - n\lambda P_n(t), \quad n \geq 1.$$

One can solve these equations recursively, but since the computations are a little messy, we will skip them and simply state that the solution is

$$P_n(t) = e^{-\lambda t}[1 - e^{-\lambda t}]^{n-1}, \quad n \geq 1.$$

(It is not too difficult to verify that $P_n(t)$ defined as above does satisfy these equations.) The form for $P_n(t)$ is nice; in fact, for a fixed t, X_t has a geometric distribution with parameter $p = e^{-\lambda t}$. This allows us immediately to compute the expected population size at time t,

$$E(X_t) = \sum_{n=1}^{\infty} nP_n(t) = e^{\lambda t}.$$

We could derive this last result in a different way. Let $f(t) = E(X_t)$. Then

$$f'(t) = \sum_{n=1}^{\infty} n P_n'(t)$$

$$= \sum_{n=1}^{\infty} n[(n-1)\lambda P_{n-1}(t) - n\lambda P_n(t)]$$

$$= \sum_{n=1}^{\infty} n\lambda P_n(t)$$

$$= \lambda f(t).$$

Therefore, $f(t)$ satisfies the standard equation for exponential growth and the initial condition $f(0) = 1$ immediately gives the solution $f(t) = e^{\lambda t}$. There is one other way we can look at the Yule process. Consider the time Y_n when the population first reaches n, i.e.,

$$Y_n = \inf\{t : X_t = n\}.$$

Then $Y_n = T_1 + \cdots + T_{n-1}$, where T_i measures the time between the arrival of the ith and $(i+1)$st individual. The random variables T_i are independent and T_i has an exponential distribution with parameter $i\lambda$. In particular $E(T_i) = 1/(i\lambda)$ and $Var(T_i) = 1/(i\lambda)^2$. Therefore,

$$E(Y_n) = \sum_{i=1}^{n-1} \frac{1}{i\lambda} \sim \frac{\ln n}{\lambda}.$$

Also $Var(Y_n) \le \sum_{i=1}^{\infty}(i\lambda)^{-2} < \infty$. Hence, Y_n equals $\ln n/\lambda$ up to a small random error which is bounded as n gets large. If it takes time $\ln n/\lambda$ to reach a population of n individuals, then in time t we would expect $e^{\lambda t}$ individuals.

Now consider the fast-growing population model, Example 5, with $\lambda_n = n^2\lambda$. Again let us consider Y_n the time until the nth individual enters the population. In this case, an interesting phenomenon occurs. Consider

$$Y_\infty = T_1 + T_2 + T_3 + \cdots.$$

Then

$$E(Y_\infty) = \sum_{i=1}^{\infty} E(T_i) = \sum_{i=1}^{\infty} \frac{1}{i^2\lambda} < \infty.$$

In particular, with probability 1, $Y_\infty < \infty$! This says that in finite time the population grows to an infinite size. This phenomenon is often called *explosion*. For a pure birth process, explosion occurs if and only if $E(Y_\infty) < \infty$, i.e., if and only if

$$\sum_{n=1}^{\infty} \lambda_n^{-1} < \infty.$$

3.4 General Case

Suppose we have a countable (perhaps infinite) state space S and rates $\alpha(x, y)$ denoting the rate at which the state is changing from x to y. Suppose for each x,

$$\alpha(x) = \sum_{y \neq x} \alpha(x, y) < \infty.$$

Then we can use the "exponential alarm clocks" at each state in order to construct a time-homogeneous, continuous-time Markov chain X_t such that for each $x \neq y$,

$$P\{X_{t+\Delta t} = y \mid X_t = x\} = \alpha(x, y)\Delta t + o_x(\Delta t).$$

Here we write $o_x(\cdot)$ to show that the size of the error term can depend on the state x. If the rates α are not bounded, it is possible for the chain to have explosion in finite time as was seen in the case of the fast-growing population model in Section 3.3. Let us assume for the time being that we have a chain for which explosion does not occur (it is sometimes difficult to determine whether or not explosion occurs).

We will consider the transition probabilities

$$p_t(x, y) = P\{X_t = y \mid X_0 = x\} = P\{X_{t+s} = y \mid X_s = x\}.$$

To derive a differential equation for the transition probabilities in the same manner as in the previous sections, we write

$$
\begin{aligned}
p_{t+\Delta t}(x, y) &= p_t(x, y)p_{\Delta t}(y, y) + \sum_{z \neq y} p_t(x, z)p_{\Delta t}(z, y) \\
&= p_t(x, y)[1 - \alpha(y)\Delta t + o_y(\Delta t)] \\
&\quad + \sum_{z \neq y} p_t(x, z)[\alpha(z, y)\Delta t + o_z(\Delta t)] \\
&= p_t(x, y)[1 - \alpha(y)\Delta t] + \sum_{z \neq y} p_t(x, z)\alpha(z, y)\Delta t \\
&\quad + \sum_z p_t(x, z)o_z(\Delta t).
\end{aligned}
$$

If we can combine the last error term so that

$$\sum_z p_t(x, z)o_z(\Delta t) = o(\Delta t), \tag{3.14}$$

then we can conclude that the transition probabilities satisfy the system of equations

$$p_t'(x, y) = -\alpha(y)p_t(x, y) + \sum_{z \neq y} \alpha(z, y)p_t(x, z),$$

where the derivative is with respect to time. These are sometimes called the *forward equations* for the chain. In most cases of interest, including all the examples in the first three sections, (3.14) can be justified. There are examples, however, where the forward equations are not correct.

There is another set of equations called the *backward equations* which always hold. For the backward equations we write

$$p_{t+\Delta t}(x,y) \;=\; \sum_{z} p_{\Delta t}(x,z)p_t(z,y)$$

$$=\; \sum_{z \neq x}[\alpha(x,z)\Delta t + o_x(\Delta t)]p_t(z,y)$$

$$+\, [1 - \alpha(x)\Delta t + o_x(\Delta t)]p_t(x,y).$$

The error term depends only on x. With a little work one can show that one can always take the limit as Δt goes to 0 and get

$$p_t'(x,y) = -\alpha(x)p_t(x,y) + \sum_{z \neq x}\alpha(x,z)p_t(z,y).$$

In the case of a finite state space with infinitesimal generator A, the backward equations for the transition matrix P_t becomes in matrix form

$$\frac{d}{dt}P_t = AP_t,$$

which can be compared to the forward equation (3.8). Both equations (with initial condition $P_0 = I$) have the solution

$$P_t = e^{At}.$$

3.5 Exercises

3.1 Suppose that the number of calls per hour arriving at an answering service follows a Poisson process with $\lambda = 4$.

(a) What is the probability that fewer than two calls come in the first hour?

(b) Suppose that six calls arrive in the first hour. What is the probability that at least two calls will arrive in the second hour?

(c) The person answering the phones waits until fifteen phone calls have arrived before going to lunch. What is the expected amount of time that the person will wait?

(d) Suppose it is known that exactly eight calls arrived in the first two hours. What is the probability that exactly five of them arrived in the first hour?

(e) Suppose it is known that exactly k calls arrived in the first four hours. What is the probability that exactly j of them arrived in the first hour?

3.2 Let X_t and Y_t be two independent Poisson processes with rate parameters λ_1 and λ_2, respectively, measuring the number of customers arriving in stores 1 and 2, respectively.

(a) What is the probability that a customer arrives in store 1 before any customers arrive in store 2?

(b) What is the probability that in the first hour, a total of exactly four customers have arrived at the two stores?

(c) Given that exactly four customers have arrived at the two stores, what is the probability that all four went to store 1?

(d) Let T denote the time of arrival of the first customer at store 2. Then X_T is the number of customers in store 1 at the time of the first customer arrival at store 2. Find the probability distribution of X_T (i.e., for each k, find $P\{X_T = k\}$).

3.3 Suppose X_t and Y_t are independent Poisson processes with parameters λ_1 and λ_2, respectively, measuring the number of calls arriving at two different phones. Let $Z_t = X_t + Y_t$.

(a) Show that Z_t is a Poisson process. What is the rate parameter for Z?

(b) What is the probability that the first call comes on the first phone?

(c) Let T denote the first time that at least one call has come from each of the two phones. Find the density and distribution function of the random variable T.

3.4 Let A be the infinitesimal generator for an irreducible, finite-state, continuous-time Markov chain. Then the rows of A add up to 0 and the nondiagonal elements of A are nonnegative.

(a) Let a be some positive number greater than all the entries of A. Let $P = (1/a)A + I$. Show that P is the transition matrix for a discrete-time, irreducible, aperiodic Markov chain.

(b) Use this to conclude: A has a unique left eigenvector with eigenvalue 0 that is a probability vector and all the other eigenvalues of A have real part strictly less than 0.

3.5 Let X_t be a Markov chain with state space $\{1, 2\}$ and rates $\alpha(1, 2) = 1, \alpha(2, 1) = 4$. Find P_t.

3.6 Let X_t be an irreducible, continuous-time Markov chain. Show that for each i, j and every $t > 0$,

$$P\{X_t = j \mid X_0 = i\} > 0.$$

3.7 Consider the continuous-time Markov chain with state space $\{1, 2, 3, 4\}$ and infinitesimal generator

$$A = \begin{bmatrix} -3 & 1 & 1 & 1 \\ 0 & -3 & 2 & 1 \\ 1 & 2 & -4 & 1 \\ 0 & 0 & 1 & -1 \end{bmatrix}.$$

(a) Find the equilibrium distribution $\bar{\pi}$.

(b) Suppose the chain starts in state 1. What is the expected amount of time until it changes state for the first time?

(c) Again assume the chain starts in state 1. What is the expected amount of time until the chain is in state 4?

3.8 Let X_t be a continuous-time birth-and-death process with birth rate $\lambda_n = 1 + (1/(n+1))$ and death rate $\mu_n = 1$. Is this process positive recurrent, null recurrent, or transient? What if $\lambda_n = 1 - (1/(n+2))$?

3.9 Consider the population model (Example 3, Section 3.3). For which values of μ and λ is extinction certain, i.e., when is the probability of reaching state 0 equal to 1?

3.10 Consider the population model with immigration (Example 4, Section 3.3). For which values of μ, λ, ν is the chain positive recurrent, null recurrent, transient?

3.11 Consider a birth-and-death process with $\lambda_n = 1/(n+1)$ and $\mu_n = 1$. Show that the process is positive recurrent and give the stationary distribution.

3.12 Suppose one has a *deterministic* model for population where the population grows proportionately to the square of the current population. In other words, the population $p(t)$ satisfies the differential equation

$$\frac{dp}{dt} = c[p(t)]^2,$$

for some constant $c > 0$. Assume $p(0) = 1$. Solve this differential equation (by separation of variables) and describe what happens as time increases.

3.13 Consider a birth-and-death process with birth rates λ_n and death rates μ_n. What are the backward equations for the transition probabilities $p_t(m, n)$?

CHAPTER 4

Optimal Stopping

4.1 Optimal Stopping of Markov Chains

Imagine the following simple game. A player rolls a die. If the player rolls a 6 the player wins no money. Otherwise, the player may either quit the game and win k dollars, where k is the roll of the die, or may roll again. If the player rolls again, the game continues until either a 6 is rolled or the player quits. The total payoff for the game is always k dollars, where k is the value of the last roll (unless the roll is a 6 in which case the payoff is 0). What is the optimal strategy for the player?

In order to determine the optimal strategy, it is necessary to decide what should be optimized. For example, if the player only wants to guarantee that the payoff is positive, then the game should be stopped after the first roll—either the player has already lost (if a 6 is rolled) or the player can guarantee a positive payoff by stopping. However, it is reasonable to consider what happens if the player decides to maximize the *expected* payoff. Let us analyze this problem and then show how this applies to more general Markov chain problems.

We first let $f(k)$ denote the payoff associated with each roll. In this example $f(k) = k$ if $k \leq 5$ and $f(6) = 0$. We let $v(k)$ be the expected winnings of the player given that the first roll is k *assuming that the player takes the optimal strategy*. At this moment we may not know what the optimal strategy is, but it still make sense to discuss v. We will, in fact, write down an equation that v satisfies and use this to determine v and the optimal strategy. We first note that $v(6) = 0$ and $v(5) = 5$. The latter is true since it clearly does not pay to roll again if the first roll is 5, so the optimal strategy is to stop and pick up \$5. It is not so clear what $v(k)$ is for $k \leq 4$.

Now let $u(k), k \leq 5$ be the amount of payoff that is expected if the player does not stop after rolling a k, but from then on plays according to the optimal strategy. [In this particular example, $u(k)$ is actually the same for all k.] Then it is easy to see that

$$u(k) = \frac{1}{6}v(1) + \frac{1}{6}v(2) + \frac{1}{6}v(3) + \frac{1}{6}v(4) + \frac{1}{6}v(5) + \frac{1}{6}v(6).$$

We now can write the optimal strategy in terms of $u(k)$—if $f(k) > u(k)$, the player should stop and take the money; if $f(k) < u(k)$, the player should roll again. In other words,

$$v(k) = \max\{f(k), u(k)\}.$$

In particular, $v(k) \geq f(k)$. This fact implies that $u(k) \geq (f(1) + \cdots + f(6))/6 = 5/2$. We now know more about the optimal strategy—if the first roll is a 1 or a 2 the player should roll again. Hence,

$$v(1) = [v(1) + \cdots + v(6)]/6 = [v(1) + \cdots + v(4)]/6 + 5/6,$$

$$v(2) = [v(1) + \cdots + v(4)]/6 + 5/6.$$

Suppose the first roll is a 4. Suppose that the optimal strategy were to continue playing. Then clearly that would also be the optimal strategy if the first roll is a 3. Under this strategy, the game would continue until a 5 or a 6 is rolled and each of these ending rolls would be equally likely. This would give an expected payoff of $(5 + 0)/2 = 5/2$, which is less than 4. Hence this cannot be the optimal strategy starting with a 4. The player, therefore, should stop with a 4 and $v(4) = f(4) = 4$. We finally consider what happens if the first roll is a 3. Suppose the player rolls again whenever a 3 comes up and uses the optimal strategy otherwise. Let u be the expected winnings in this case. Then

$$u = P\{\text{roll} \leq 3\}u + (1/6)4 + (1/6)5 = (1/2)u + (1/6)4 + (1/6)5.$$

Solving for u we get $u = 9/3$. Since this equals $f(3)$, the expected payoff for playing is the same as for stopping and $v(3) = 3$. With these values, we can solve for $v(1)$ and $v(2)$, getting $v(1) = v(2) = 3$. The optimal strategy is to play if the first roll is 1 or 2; stop if the first roll is $4, 5, 6$; and either play or stop if the first roll is a 3.

We now generalize these ideas. Suppose P is the transition matrix for a discrete-time Markov chain X_n with state space S. For ease we will assume that S is finite, but much of what follows can be applied to the infinite state space case. Assume there is a payoff function f that assigns to each state the payoff if the chain is stopped when it reaches that state. In cases of interest, P will not be irreducible since otherwise one could always continue until one reached the state that has the maximum payoff. A *stopping rule* or *stopping time* will be a random variable T that gives the time at which the chain is stopped. It is important that one must decide whether or not to stop based only on what has happened up through step n; in other words, one cannot look into the future to decide whether or not to stop. Because we are dealing with a time-homogeneous Markov chain it does not take too much work to convince oneself that the only reasonable stopping rules that do not look into the future are of the following form: the state space is divided into two sets S_1 and S_2; if the state of the chain is in S_1 one continues, if it is in S_2 it stops. The goal is to maximize the expected payoff over all stopping rules. We let $v(x)$ be the *value* of a state x, i.e., the expected payoff assuming that the optimal stopping strategy is used. We can write

$$v(x) = \sup_T E(f(X_T) \mid X_0 = x),$$

where the supremum is over all legal stopping rules.

There are two main inequalities that v satisfies. First, v is greater than or equal to the payoff available by stopping,

$$v(x) \geq f(x). \tag{4.1}$$

Second, v is greater than or equal to the maximum expected payoff if one continues,

$$v(x) \geq Pv(x) = \sum_{y \in S} p(x, y)v(y). \tag{4.2}$$

In fact, v is equal to the maximum of these values:

$$v(x) = \max\{f(x), Pv(x)\}. \tag{4.3}$$

If we let S_1 be the set of states where one continues and S_2 the set of states where one stops (assuming the optimal strategy), and we let

$$T = \min\{j \geq 0 : X_j \in S_2\},$$

then

$$v(x) = E(f(X_T) \mid X_0 = x).$$

We will characterize the function v. We call a function u *superharmonic* with respect to P if it satisfies (4.2), i.e.,

$$u(x) \geq Pu(x).$$

Suppose u is superharmonic and T is the time associated to a stopping rule as above. Consider the time $T_n = \min\{T, n\}$ We claim that

$$u(x) \geq E(u(T_n) \mid X_0 = x).$$

To see this, note that it is trivially true for $n = 0$. Assume it is true for $n - 1$. Then

$$E(u(T_n) \mid X_0 = x)$$

$$= \sum_{y \in S} P\{X_{T_n} = y \mid X_0 = x\}u(y)$$

$$= \sum_{y \in S}\sum_{z \in S} P\{X_{T_n} = y \mid X_{T_{n-1}} = z\}P\{X_{T_{n-1}} = z \mid X_0 = x\}u(y)$$

$$= \sum_{z \in S_2}\sum_{y \in S} P\{X_{T_n} = y \mid X_{T_{n-1}} = z\}P\{X_{T_{n-1}} = z \mid X_0 = x\}u(y) +$$

$$\sum_{z \in S_1}\sum_{y \in S} P\{X_{T_n} = y \mid X_{T_{n-1}} = z\}P\{X_{T_{n-1}} = z \mid X_0 = x\}u(y).$$

If $z \in S_2$, then $P\{X_{T_n} = z \mid X_{T_{n-1}} = z\} = 1$ and hence the first double sum in the last expression equals

$$\sum_{z \in S_2} P\{X_{T_{n-1}} = z \mid X_0 = x\} u(z).$$

If $z \in S_1$, $P\{X_{T_n} = y \mid X_{T_{n-1}} = z\} = p(z, y)$ and hence

$$\sum_{y \in S} P\{X_{T_n} = y \mid X_{T_{n-1}} = z\} u(y) = P u(z) \leq u(z).$$

Hence,

$$
\begin{aligned}
E(u(T_n) \mid X_0 = x\} &\leq \sum_{z \in S} P\{X_{T_{n-1}} = z \mid X_0 = x\} u(z) \\
&= E(u(T_{n-1}) \mid X_0 = x) \leq u(x).
\end{aligned}
$$

Since u is a bounded function, we can let $n \to \infty$ and get

$$u(x) \geq \lim_{n \to \infty} E(u(T_n) \mid X_0 = x) = E(u(T) \mid X_0 = x).$$

Now suppose that $u(x) \geq f(x)$ for all x. Then

$$u(x) = E(u(T) \mid X_0 = x) \geq E(f(T) \mid X_0 = x) = v(x).$$

Hence every superharmonic function that is larger than f is greater than or equal to the value function v. Also we note (see Exercise 4.6) that if $\{u_i(x)\}$ is any collection of superharmonic functions, then

$$u(x) = \inf_i u_i(x)$$

is also superharmonic. We therefore have the following: v is the smallest superharmonic function with respect to P that is greater than equal to f; equivalently,

$$v(x) = \inf u(x),$$

where the infimum is over all superharmonic functions u with $u(x) \geq f(x)$.

The characterization leads to an algorithm for determining v. Start with the function $u_1(x)$ that equals $f(x)$ if x is an absorbing state and otherwise equals the maximum value of f. This gives a superharmonic function that is greater than f. Let

$$u_2(x) = \max\{P u_1(x), f(x)\}.$$

Since u_1 is superharmonic and $u_1 \geq f$, $u_2(x) \leq u_1(x)$. Also,

$$P u_2(x) \leq P u_1(x) \leq u_2(x).$$

Hence, u_2 is a superharmonic function greater than f. Continuing, we define

$$u_n(x) = \max\{P u_{n-1}(x), f(x)\},$$

and we see that u_n is a superharmonic function greater than f but less than u_{n-1}. We can then check (see the end of this section) that

$$v(x) = \lim_{n \to \infty} u_n(x).$$

Example 1. If we consider the game that we already analyzed and started with the function $u = [5, 5, 5, 5, 5, 0]$, then in 10 iterations we would see $u_{10} = [3.002, 3.002, 3.002, 4, 5, 0]$.

Example 2. Suppose X_n is a simple random walk ($p = 1/2$) with absorbing barriers on $\{0, 1, 2, 3, 4, 5, 6\}$. Let the payoff function f be given by $f = [0, 2, 4, 5, 9, 3, 0]$ (we write the payoff function as a vector in a natural way). We will first determine the optimal strategy. Clearly one stops at state 4 and one has to stop at 0 and 6. From state 5 there is a probability $1/2$ of going to 4 and $1/2$ of going to 6; the expected payoff given that we continue is at least $9/2 > f(5) = 3$, so from 5 we continue. If one starts in state 3, then one can get an expected payoff of $(4+9)/2 = 13/2$ by taking one step and then stopping. Since this is greater than $f(3) = 5$, it must be optimal to continue from state 3 and $v(3) \geq 13/2$. Note that from state 2 one can get by playing an expected payoff of at least $[f(1) + v(3)]/2 \geq 17/4 \geq f(2) = 4$. Hence, we continue on state 2 and $v(2) \geq 17/4$. Similarly, if one continues from state 1 we can obtain an expected payoff of $v(2)/2 \geq 17/8 > f(1) = 2$, so the optimal strategy is to continue. Therefore the stopping set in this case is $S_2 = \{0, 4, 6\}$. The value function can be obtained by

$$v(x) = E(f(X_T) \mid X_0 = x) = 9P\{X_T = 4 \mid X_0 = x\}.$$

The corresponding probability has been computed before [see (1.16)] and we get

$$v = [0, 9/4, 9/2, 27/4, 9, 9/2, 0].$$

In this example, if we had started with the function $u_1 = [0, 9, 9, 9, 9, 9, 0]$ and performed the algorithm above we would have gotten within .01 of the actual value of v in about 20 iterations.

In solving the optimal stopping problem we simultaneously compute the value function v and the optimal stopping strategy. Suppose that we knew the strategy that we would choose, i.e., we split the state space into two sets S_1 and S_2 so that we continue on S_1 and stop on S_2. Let $u(x)$ be the expected payoff using this strategy. Then u satisfies:

$$u(x) = f(x), \quad x \in S_2, \tag{4.4}$$

$$u(x) = Pu(x), \quad x \in S_1. \tag{4.5}$$

This is a discrete analogue of a boundary value problem sometimes called the Dirichlet problem. The boundary is the set S_2 where prescribed values are given. On the "interior" points S_1, some difference equation holds. As we have seen the probabilistic form of the solution of this system can be given by

$$u(x) = E(f(X_T) \mid X_0 = x),$$

where

$$T = \min\{j \geq 0 : X_j \in S_2\}.$$

For a finite-state Markov chain, the solution can be found directly because (4.4) and (4.5) give essentially k linear equations in k unknowns, where k is the number of points in S_1 and the unknowns are $u(x), x \in S_1$.

We now verify that the algorithm does converge to the value function v. Let $u(z) = \lim_{n \to \infty} u_n(z)$. Since u is the decreasing limit of superharmonic functions, u is superharmonic (see Exercise 4.6). Also $u(z) \geq f(z)$ for all z. Hence by the characterization of v, we get

$$u(z) \geq v(z). \tag{4.6}$$

Let the stopping set S_2 be defined by

$$S_2 = \{z : u(z) = f(z)\},$$

$$S_1 = \{z : u(z) > f(z)\}.$$

On S_1, $Pu(z) = u(z)$ (if $Pu(z) < u(z)$, then for some n, $Pu_n(z) < u(z) \leq u_n(z)$ and hence $u_{n+1}(z) = \max\{Pu_n(z), f(z)\} < u(z)$ which is impossible). Therefore,

$$u(z) = E(f(X_T) \mid X_0 = z),$$

where T is the strategy associated with the sets S_1, S_2. Since $v(z)$ is the largest expected value over all choices of stopping sets,

$$u(z) \leq v(z). \tag{4.7}$$

Combining (4.6) and (4.7) we see that $u(z) = v(z)$ for all z.

4.2 Optimal Stopping with Cost

Consider the first example and suppose that there is a charge of $1 for each additional roll, i.e., on each roll we can either take the payoff associated with that roll or pay $1 and roll again. In general, we may assume that there is a cost $g(x)$ associated with each state that must be paid to continue the chain. Again we assume we have a payoff function f and we let $v(x)$ be the expected value of the payoff *minus the cost* assuming a stopping rule is chosen that maximizes this expected value. We can write

$$v(x) = \sup_{T} E\left(f(X_T) - \sum_{j=0}^{T-1} g(X_j) \mid X_0 = x\right),$$

where again the supremum is over all legal stopping times T. Then $v(x)$ satisfies:

$$v(x) = \max\{f(x), Pv(x) - g(x)\}.$$

Here, the expected payoff minus cost if the chain is continued is $Pv(x) - g(x)$. Again we can divide S into S_1 and S_2 where

$$S_2 = \{x : v(x) = f(x)\},$$

and the optimal stopping rule is to stop when the chain enters a state in S_2.

Using a similar argument as in Section 4.1, the value function v for this example can be characterized as the smallest function u greater than f that satisfies

$$u(x) \geq Pu(x) - g(x).$$

In other words,

$$v(x) = \inf u(x),$$

where the infimum is over all u satisfying $u(x) \geq f(x)$ and $u(x) \geq Pu(x) - g(x)$. To find the value function, we may use an algorithm similar to that in Section 4.1. We define u_1 to be the function that equals f on all absorbing states and equals the maximum value of f everywhere else. We then define

$$u_n(x) = \max\{f(x), Pu_{n-1}(x) - g(x)\},$$

and then

$$v(x) = \lim_{n \to \infty} u_n(x).$$

Example 1. Suppose we consider the die game with $f = [1, 2, 3, 4, 5, 0]$ and $g = [1, 1, 1, 1, 1, 1]$. The cost function makes it less likely that we would want to roll again, so it is clear that we should stop if we get a 4 or a 5; similarly, we should stop if we get a 3 since we were indifferent before with this roll and it costs to roll again. If we get a 1, then by rolling again we can get an expected payoff of at least 5/2 with a cost of 1. Hence we can expect a net gain of at least 3/2. Therefore we should play if we get a 1.

Suppose we roll again whenever we get a 1 or 2 and stop otherwise. Let $u(k)$ be the expected winnings with this strategy. Then $u(1) = u(2) = u$ and $u(k) = k, k = 3, 4, 5$. Also,

$$u(2) = \frac{1}{6}u(1) + \frac{1}{6}u(2) + \frac{1}{6}u(3) + \frac{1}{6}u(4) + \frac{1}{6}u(5) + \frac{1}{6}u(6) - 1 = \frac{1}{3}u + 1.$$

Solving for u gives $u = 3/2$. Since this is less than $f(2) = 2$, it must be correct to stop at 2. Hence the stopping set is $S_2 = \{2, 3, 4, 5, 6\}$ and the value function is

$$v = [8/5, 2, 3, 4, 5, 0].$$

If we started with the initial $u_1 = [5, 5, 5, 5, 5, 0]$ and performed the algorithm described above, then after only a few iterations we would have

$$u_{10} = [1.6, 2, 3, 4, 5, 0].$$

Example 2. Consider the other example of the previous section where X_n is a simple random walk with absorbing boundary on $\{0, 1, \ldots, 6\}$ and $f = [0, 2, 4, 5, 9, 3, 0]$. Suppose we impose a cost of .5 to move from states $0, 1, 2$ and a cost of 1 to move from $3, 4, 5, 6$, i.e., a cost function

$$g = [.5, .5, .5, 1, 1, 1, 1].$$

If we start with initial $u_1 = [0, 9, 9, 9, 9, 9, 0]$, then in only six iterations we get

$$u_6 = [0, 2, 4, 5.5, 9, 3.5, 0],$$

which gives the value for v. In this case the stopping set is $S_2 = \{0, 1, 2, 4, 6\}$.

Example 3. With a cost function, it is possible to have a nontrivial problem even if the Markov chain is irreducible. Suppose we play the following game: roll two dice; the player may stop at any time and take the roll on the dice or the player may pay 2 units if the roll is less than 5 and 1 unit if the roll is greater than or equal to 5 and roll again. In this case the state space is $\{2, 3, 4, \ldots, 12\}$,

$$f = [2, 3, 4, \ldots, 12], \quad g = [2, 2, 2, 1, 1, \ldots, 1].$$

If we start with the initial guess $u_1 = [12, 12, \ldots, 12]$ then within 20 iterations we converge to the value function v,

$$v = [5\frac{2}{3}, 5\frac{2}{3}, 5\frac{2}{3}, 6\frac{2}{3}, 6\frac{2}{3}, 7, 8, 9, 10, 11, 12].$$

The stopping set is $S_2 = \{7, 8, \ldots, 12\}$.

4.3 Optimal Stopping with Discounting

It is often appropriate in modelling financial matters to assume that the value of money decreases with time. Let us assume that a discount factor $\alpha < 1$ is given. By this we mean that 1 dollar received after one time unit is the the same as α dollars received in the present. Again suppose we have a Markov chain X_n with transition matrix P and a payoff function f. It is now the goal to optimize the expected value of the payoff, taking into consideration the decreasing value of the payoff. If we stop after k steps, then the present value of the payoff in k steps is α^k time the actual payoff.

In this case the value function is given by

$$v(x) = \sup_T E(\alpha^T f(X_T) \mid X_0 = x),$$

where again the supremum is over all legal stopping rules. To obtain this value function, we characterize v as the smallest function u satisfying

$$u(x) \geq f(x),$$

$$u(x) \geq \alpha P u(x).$$

We may obtain v with a similar algorithm as before. Start with an initial function u_1 equal to f at all absorbing states and equal to the maximum value of f at all other states. Then define u_n recursively by

$$u_n(x) = \max\{f(x), \alpha P u_{n-1}(x)\}.$$

Then

$$v(x) = \lim_{n \to \infty} u_n(x).$$

Example 1. Consider the die game again. Assume a discounting factor of $\alpha = .8$. Since discounting can only make it more likely to stop it is easy to see that one should stop if the first roll is a 3, 4, or 5. If the first roll is a 1, one can get an expected payoff of at least $.8[(1+2+3+4+5)/6] = 2$ by rolling again, so it is best to roll again. Suppose we use the strategy to roll again with a 1, 2 and to stop otherwise and let u be the expected winnings given that one rolls again. Then

$$u = .8[u/6 + u/6 + 3/6 + 4/6 + 5/6].$$

Solving for u we get $u = 24/11 > 2$ so it must be optimal to roll again with a 2. Therefore $S_2 = \{3, 4, 5, 6\}$ and

$$v = [24/11, 24/11, 3, 4, 5, 0].$$

Example 2. Consider the example of a simple random walk with absorbing boundaries on $\{0, 1, \ldots, 6\}$ and $f = [0, 2, 4, 5, 9, 3, 0]$. Suppose that there is no cost function, but the value of money is discounted at rate $\alpha = .9$. If we start with $u_1 = [0, 9, 9, 9, 9, 9, 0]$ then in seven iterations we converge to the value

$$u_7 = [0, 2, 4, 5.85, 9, 4.05, 0].$$

This stopping set is $\{0, 1, 2, 4, 6\}$.

It is possible to include both a cost function and a discounting factor. Suppose in addition to the other assumptions in this section, we have a cost function $g(x)$ that indicates the cost of taking a step given that the chain is in state x. Then the value function v is the smallest function u satisfying

$$u(x) \geq f(x),$$
$$u(x) \geq \alpha P u(x) - g(x),$$

Example 3. Consider the random walk with absorbing boundaries described before with $f = [0, 2, 4, 5, 9, 3, 0]$ and with both the cost function $g = [.5, .5, .5, 1, 1, 1, 1]$ and the discount factor $\alpha = .9$. If we start with $u_1 = [0, 9, 9, 9, 9, 9, 0]$ then in only three iterations we converge to

$$v = [0, 2, 4, 5, 9, 3.05, 0].$$

The stopping set is $\{0, 1, 2, 3, 4, 6\}$.

Example 4. Consider a random walk with absorbing boundaries on the state space $\{0, 1, \ldots, 10\}$. Suppose the payoff function is the square of the site stopped at, i.e.,

$$f = [0, 1, 4, 4, 9, \ldots, 100].$$

We assume that there is a constant cost of .6 and a discounting factor of $\alpha = .95$. We then start with the initial guess

$$u_1 = [0, 100, 100, 100, \ldots, 100]$$

and after 60 iterations we get

$$u_{60} = [0, 1.51, 4.45, 9.11, 16, 25, 36, 49, 64, 81, 100].$$

The stopping set is $\{0, 4, 5, 6, \ldots, 10\}$.

4.4 Exercises

4.1 Consider a simple random walk ($p = 1/2$) with absorbing boundaries on $\{0, 1, 2, \ldots, 10\}$. Suppose the following payoff function is given

$$[0, 2, 4, 3, 10, 0, 6, 4, 3, 3, 0].$$

Find the optimal stopping rule and give the expected payoff starting at each site.

4.2 The following game is played: you roll two dice. If you roll a 7, the game is over and you win nothing. Otherwise, you may stop and receive an amount equal to the sum of the two dice. If you continue, you roll again. The game ends whenever you roll a 7 or whenever you say stop. If you say stop before rolling a 7 you receive an amount equal to the sum of the two dice on the last roll. What is your expected winnings: a) if you always stop after the first roll; b) if you play to optimize your expected winnings?

4.3 Consider Exercise 4.1. Do the problem again assuming:
  (a) a constant cost of .75 for each move;
  (b) a discount factor $\alpha = .95$;
  (c) both.

4.4 Consider Exercise 4.2. Do the problem again assuming:
  (a) a cost function of $g = [2, 2, 2, 2, 1, 1, 1, 1, 1, 1, 1]$;
  (b) a discount factor $\alpha = .8$;
  (c) both.

4.5 Consider a simple random walk on the following four-vertex graph.

Assume that the payoff function is: $f(A) = 2, f(B) = 4, f(C) = 5, f(D) = 3$. Assume that there is no cost associated with moving, but there is a discount factor α. What is the largest possible value of α so that the optimal stopping strategy is to stop at every vertex, i.e., so that $S_2 = \{A, B, C, D\}$?

4.6 If $u_1(y), u_2(y), \ldots$ are all functions that are superharmonic at x for P, i.e.,

$$Pu_i(x) \leq u_i(x),$$

and we let u be the function

$$u(y) = \inf_i u_i(y),$$

show that u is superharmonic at x for P.

4.7 Consider a simple "Wheel of Fortune" game. A wheel is divided into 12 equal-sized wedges. Eleven of the edges are marked with the numbers $100, 200, \ldots, 1100$ denoting an amount of money won if the wheel lands on those numbers. The twelfth wedge is marked "bankrupt." A player can spin as many times as he or she wants. Each time the wheel lands on a numbered wedge, the player receives that much money which is added to his/her previous winnings. However, if the wheel ever lands on the "bankrupt" wedge, the player lose all of his/her money that has been won up to that point. The player may quit at any time, and take all the money he or she has won (assuming the "bankrupt" wedge has not come up).

Assuming that the goal is to maximize one's expected winnings in this game, devise an optimal strategy for playing this game and compute one's expected winnings. You may wish to try a computer simulation first.

CHAPTER 5

Martingales

5.1 Conditional Expectation

In this section we discuss the notion of conditional expectation. We start with some easier examples and build up to a general definition. Suppose Y is a random variable measuring the outcome of some random experiment. If one knows nothing about the outcome of the experiment, then the best guess for the value of Y is $E(Y)$, the expectation. Of course, if one has complete knowledge of the outcome of the experiment, then one knows the exact value of Y. The notion of conditional expectation deals with making the best guess for Y given some but not complete information about the outcome. To give a general definition of conditional expectation one needs the framework of measure theoretic probability theory. We will not aim for a general definition, but instead restrict ourselves to discussing the conditional expectation of a random variable Y with respect to a finite number of random variables $X_1, \ldots, X_n$.

Suppose that X and Y are discrete random variables with joint probability density function

$$f(x, y) = P\{X = x, Y = y\}$$

and marginal probability density functions

$$f_X(x) = \sum_y f(x, y), \quad f_Y(y) = \sum_x f(x, y).$$

To define the conditional expectation of Y given X, $E(Y \mid X)$ we need to give the best value of Y for any value of x. A little thought will show that we should define

$$
\begin{aligned}
E(Y \mid X)(x) &= \sum_x y P\{Y = y \mid X = x\} \\
&= \sum_x y P\{X = x, Y = y\} / P\{X = x\} \\
&= [\sum_x y f(x, y)] / f_X(x).
\end{aligned}
$$

This is well-defined if $f_X(x) > 0$ and we do not worry about defining $E(Y \mid X)(x)$ for other values of x since such values occur with probability 0. As an example suppose that two independent dice are rolled and X denotes the value of the first roll and Y denotes the sum of the two rolls. Then

$$f(x,y) = 1/36, \quad x = 1, 2, \ldots 6, \ y = x + 1, x + 2, \ldots x + 6,$$

and

$$E(Y \mid X)(x) = x + \frac{7}{2}.$$

Similarly, if $X_1, \ldots, X_n, Y$ are discrete random variables with joint probability density function

$$f(x_1, \ldots, x_n, y) = P\{X_1 = x_1, \ldots, X_n = x_n, Y = y\},$$

and the marginal density with respect to $X_1, \ldots, X_n$ is given by

$$g(x_1, \ldots, x_n) = \sum_y f(x_1, \ldots, x_n, y),$$

then the conditional expectation of Y given $X_1, \ldots, X_n$, is given by

$$E(Y \mid X_1, \ldots, X_n)(x_1, \ldots, x_n) = [\sum_y yf(x_1, \ldots, x_n, y)]/g(x_1, \ldots, x_n).$$

This is well-defined if $x_1, \ldots, x_n$ is a possible outcome for the experiment, i.e., if $g(x_1, \ldots, x_n) > 0$. Again, we think of $E(Y \mid X_1, \ldots X_n)$ as being the best guess for the value of Y given the values of $X_1, \ldots, X_n$.

If X and Y are continuous random variables with joint density $f(x,y)$ and marginal densities

$$f_X(x) = \int_{-\infty}^{\infty} f(x,y)dy, \quad f_Y(y) = \int_{-\infty}^{\infty} f(x,y)dx,$$

then the conditional expectation of Y given X is defined in an analogous way

$$E(Y \mid X)(x) = [\int_{-\infty}^{\infty} yf(x,y)dy]/f_X(x),$$

which is well-defined for $f_X(x) > 0$. Similarly if $X_1, \ldots X_n, Y$ have joint density $f(x_1, \ldots, x_n, y)$,

$$E(Y \mid X_1, \ldots, X_n)(x_1, \ldots, x_n) =$$

$$[\int_{-\infty}^{\infty} yf(x_1, \ldots, x_n, y)dy]/f_{X_1, \ldots, X_n}(x_1, \ldots, x_n).$$

The conditional expectation $E(Y \mid X_1, \ldots, X_n)$ is characterized by two properties:

1. The value of the random variable $E(Y \mid X_1, \ldots, X_n)$ depends only on the values of $X_1, \ldots, X_n$, i.e., we can write $E(Y \mid X_1, \ldots, X_n) =$

$\phi(X_1,\ldots,X_n)$ for some function ϕ (for those who know measure theory, the function ϕ must be a measurable function). If a random variable Z can be written as a function of $X_1,\ldots,X_n$ it is called *measurable* with respect to $X_1,\ldots,X_n$.

2. Suppose A is any event that depends only on $X_1,\ldots,X_n$. For example, A might be the event

$$A = \{a_1 \leq X_1 \leq b_1, \ldots, a_n \leq X_n \leq b_n\}.$$

Let I_A denote the indicator function of A, i.e., the random variable which equals 1 if A occurs and 0 otherwise. Then

$$E(YI_A) = E(E(Y \mid X_1,\ldots,X_n)I_A). \tag{5.1}$$

Let us illustrate the last equality in the case where $X_1,\ldots,X_n,Y$ are continuous random variables with density $f(x_1,\ldots,x_n,y)$ and A is the above event. In this case,

$$
\begin{aligned}
&E(E(Y \mid X_1,\ldots,X_n)I_A) \\[4pt]
&= \int_{a_1}^{b_1} \cdots \int_{a_n}^{b_n} \int_{-\infty}^{\infty} E(Y \mid X_1 = x_1,\ldots,X_n = x_n) \\
&\hspace{6cm} f(x_1,\ldots,x_n,y)\, dy\, dx_n \cdots dx_1 \\[4pt]
&= \int_{a_1}^{b_1} \cdots \int_{a_n}^{b_n} \int_{-\infty}^{\infty} \left[\frac{\int_{-\infty}^{\infty} z f(x_1,\ldots,x_n,z)\, dz}{\int_{-\infty}^{\infty} f(x_1,\ldots,x_n,z)\, dz}\right] \\
&\hspace{6cm} f(x_1,\ldots,x_n,y)\, dy\, dx_n \cdots dx_1 \\[4pt]
&= \int_{a_1}^{b_1} \cdots \int_{a_n}^{b_n} \int_{-\infty}^{\infty} z f(x_1,\ldots,x_n,z)\, dz\, dx_n \cdots dx_1 \\[4pt]
&= E(YI_A).
\end{aligned}
$$

Conditions 1 and 2 give a complete characterization of the conditional expectation $E(Y \mid X_1,\ldots X_n)$: it is the unique random variable which depends only on $X_1,\ldots,X_n$ and which satisfies (5.1) for every event A that depends only on $X_1,\ldots,X_n$. In a measure theoretic treatment of probability, the conditional expectation would be defined as the random variable satisfying conditions 1 and 2 and then it would be proved that this uniquely defines a random variable (up to an event of probability 0). For our purposes, the characterization will be useful in deriving some properties of conditional expectation.

We will make the notation a little more compact. If $X_1, X_2, \ldots$ is a sequence of random variables we will use $\mathcal{F}_n$ to denote the "information contained in $X_1,\ldots,X_n$." We will write $E(Y \mid \mathcal{F}_n)$ for $E(Y \mid X_1,\ldots,X_n)$. If we apply (5.1) to the event A consisting of the entire sample space (so that $I_A = 1$) we get

$$E(E(Y \mid \mathcal{F}_n)) = E(Y). \tag{5.2}$$

Conditional expectation is a linear operation; if a, b are constants

$$E(aY_1 + bY_2 \mid \mathcal{F}_n) = aE(Y_1 \mid \mathcal{F}_n) + bE(Y_2 \mid \mathcal{F}_n). \qquad (5.3)$$

To prove this, we need only note that the right-hand side is measurable with respect to $X_1, \ldots, X_n$ and satisfies (5.1). The next two properties can be derived similarly. If Y is already a function of $X_1, \ldots, X_n$ then

$$E(Y \mid \mathcal{F}_n) = Y. \qquad (5.4)$$

For any Y, if $m < n$, then

$$E(E(Y \mid \mathcal{F}_n) \mid \mathcal{F}_m) = E(Y \mid \mathcal{F}_m). \qquad (5.5)$$

If Y is independent of $X_1, \ldots, X_n$, then information about $X_1, \ldots, X_n$ should not be useful in determining Y and

$$E(Y \mid \mathcal{F}_n) = E(Y). \qquad (5.6)$$

This can be derived easily from (5.1) since in this case Y and I_A are independent random variables. The last property we will need is a little trickier: if Y is any random variable and Z is a random variable that is measurable with respect to $X_1, \ldots X_n$, then

$$E(YZ \mid \mathcal{F}_n) = ZE(Y \mid \mathcal{F}_n). \qquad (5.7)$$

It is clear that the right-hand side is measurable with respect to $X_1, \ldots, X_n$, so it suffices to show that it satisfies (5.1). We will not prove it here; the basic idea is to approximate Z by simple functions, for which (5.1) can be derived easily, and pass to the limit.

Example 1. Suppose $X_1, X_2, \ldots$ are independent, identically distributed random variables with mean μ. Let S_n denote the partial sum

$$S_n = X_1 + \cdots + X_n.$$

Let $\mathcal{F}_n$ denote the information in $X_1, \ldots, X_n$. Suppose $m < n$. Then by (5.3),

$$E(S_n \mid \mathcal{F}_m) = E(X_1 + \cdots + X_m \mid \mathcal{F}_m) + E(X_{m+1} + \cdots + X_n \mid \mathcal{F}_m).$$

Since $X_1 + \cdots + X_m$ is measurable with respect to $X_1, \ldots, X_m$, by (5.4),

$$E(X_1 + \cdots + X_m \mid \mathcal{F}_m) = X_1 + \cdots + X_m = S_m.$$

Since $X_{m+1} + \cdots + X_n$ is independent of $X_1, \ldots, X_m$, by (5.6),

$$E(X_{m+1} + \cdots + X_n \mid \mathcal{F}_m) = E(X_{m+1} + \cdots + X_n) = (n - m)\mu.$$

Therefore

$$E(S_n \mid \mathcal{F}_m) = S_m + (n - m)\mu. \qquad (5.8)$$

Example 2. Suppose $X_1, X_2, \ldots$ and S_n are as in Example 1. Suppose

$\mu = 0$ and $Var(X_i) = E(X_i^2) = \sigma^2$. Let $m < n$. Then, by (5.3),

$$\begin{aligned} E(S_n^2 \mid \mathcal{F}_m) &= E([S_m + (S_n - S_m)]^2 \mid \mathcal{F}_m) \\ &= E(S_m^2 \mid \mathcal{F}_m) + 2E(S_m(S_n - S_m) \mid \mathcal{F}_m) \\ &\quad + E((S_n - S_m)^2 \mid \mathcal{F}_m). \end{aligned}$$

Since S_m depends only on $X_1, \ldots, X_m$ and $S_n - S_m$ is independent of $X_1, \ldots, X_m$, we have as in the previous example

$$E(S_m^2 \mid \mathcal{F}_m) = S_m^2,$$

$$E((S_n - S_m)^2 \mid \mathcal{F}_m) = E((S_n - S_m)^2) = Var(S_n - S_m) = (n - m)\sigma^2.$$

Finally, by (5.7),

$$E(S_m(S_n - S_m) \mid \mathcal{F}_m) = S_m E(S_n - S_m \mid \mathcal{F}_m) = S_m E(S_n - S_m) = 0.$$

Therefore,

$$E(S_n^2 \mid \mathcal{F}_m) = S_m^2 + (n - m)\sigma^2.$$

Example 3. Consider a special case of Example 1 where the random variable X_i has a Bernoulli distribution, $P\{X_i = 1\} = p$, $P\{X_i = 0\} = 1 - p$. Again assume that $m < n$. For any $i \leq m$, consider $E(X_i \mid S_n)$. If $S_n = k$ then there are k 1s in the first n trials. Given $S_n = k$ it is an easy exercise in conditional probability to show that $P\{X_i = 1 \mid S_n = k\} = k/n$. Hence,

$$E(X_i \mid S_n) = S_n/n, \qquad = 1 \cdot P(X_i = 1 \mid S_n = k + 0 \cdot P(x_i = 0 \mid S_n = k)$$

and

$$E(S_m \mid S_n) = E(X_1 \mid S_n) + \cdots + E(X_m \mid S_n) = (m/n)S_n.$$
$$\underbrace{\qquad\qquad}_{S \cdot Y_n \cdot}$$

In Chapters 8 and 9 we will need to consider conditional expectations with respect to an infinite collection of random variables,

$$E(Y \mid X_\alpha, \alpha \in \mathcal{A}).$$

With the aid of measure theory, one can define this as the unique random variable Z which is measurable with respect to $\{X_\alpha : \alpha \in \mathcal{A}\}$ (i.e., if you know the values of all the random variables X_α, you know the value of Z) and such that for any event A depending only on $\{X_\alpha : \alpha \in \mathcal{A}\}$,

$$E(YI_A) = E(ZI_A).$$

All of the properties (5.2) – (5.7) hold for this more general conditional expectation.

5.2 Definition and Examples

A martingale is a model of a fair game. Let $X_0, X_1, \ldots$ be a sequence of random variables. We again let $\mathcal{F}_n$ denote the information contained in

$X_0, \ldots, X_n$. We say that a sequence of random variables $M_0, M_1, M_2, \ldots$ with $E(|M_i|) < \infty$ is a *martingale* with respect to $\mathcal{F}_n$ if: 1) each M_n is measurable with respect to $X_0, \ldots, X_n$; and 2) for each $m < n$,

$$E(M_n \mid \mathcal{F}_m) = M_m, \tag{5.9}$$

or equivalently,

$$E(M_n - M_m \mid \mathcal{F}_m) = 0.$$

The condition $E(|M_i|) < \infty$ is needed to guarantee that the conditional expectations are well defined. Sometimes we will just say $M_0, M_1, \ldots$ is a martingale without making reference to the random variables $X_0, X_1, \ldots$. In this case it will mean that the sequence M_n is a martingale with respect to itself, i.e., where $\mathcal{F}_n$ is the information obtained from $M_0, \ldots, M_n$ (in which case condition 1 is trivially true). In order to verify (5.9) it suffices to prove that for all n,

$$E(M_{n+1} \mid \mathcal{F}_n) = M_n,$$

since if this holds, by (5.5),

$$
\begin{aligned}
E(M_{n+2} \mid \mathcal{F}_n) &= E(E(M_{n+2} \mid \mathcal{F}_{n+1}) \mid \mathcal{F}_n) \\
&= E(M_{n+1} \mid \mathcal{F}_n) = M_n,
\end{aligned}
$$

and so on.

Example 1. Let $X_1, X_2, \ldots$ be independent random variables each with mean μ. Let $S_0 = 0$ and for $n > 0$ let S_n be the partial sum

$$S_n = X_1 + \cdots + X_n.$$

Then $M_n = S_n - n\mu$ is a martingale with respect to $\mathcal{F}_n$, the information contained in $X_0, \ldots, X_n$. This can easily be checked by using Example 1 of Section 5.1,

$$E(M_{n+1} \mid \mathcal{F}_n) = E(S_{n+1} - (n+1)\mu \mid \mathcal{F}_n) = E(S_{n+1} \mid \mathcal{F}_n) - (n+1)\mu$$
$$= (S_n + \mu) - (n+1)\mu = M_n.$$

In particular, if $\mu = 0$, then S_n is a martingale with respect to $\mathcal{F}_n$.

Example 2. The following is a version of the "martingale betting strategy" which is a way to beat a fair game. Suppose $X_1, X_2, \ldots$ are independent random variables with

$$P\{X_i = 1\} = P\{X_i = -1\} = 1/2.$$

We can think of the random variables X_i as the results of a game where one flips a coin and wins \$1 if it comes up heads and loses \$1 if it comes up tails. One way to beat the game is to keep doubling our bet until we eventually win. At this point we stop. Let W_n denote the winnings (or losses) up

Fair games are all martingale (handwritten)

through n flips of the coin using this strategy. $W_0 = 0$. Whenever we win we stop playing, so our winnings stop changing and

$$P\{W_{n+1} = 1 \mid W_n = 1\} = 1.$$ *if I win* (handwritten)

Now suppose the first n flips of the coin have turned up tails. After each flip we have doubled our bet, so we have lost $1 + 2 + 4 + \cdots + 2^{n-1} = 2^n - 1$ dollars and $W_n = -(2^n - 1)$. At this time we double our bet again and wager 2^n on the next flip. This gives

this is what I would have to wager as I loose (handwritten)

$$P\{W_{n+1} = 1 \mid W_n = -(2^n - 1)\} = 1/2,$$

cumulative loss (handwritten) $$P\{W_{n+1} = -(2^{n+1} - 1) \mid W_n = -(2^n - 1)\} = 1/2.$$

$1 + 2 + 4 + 8 + \cdots$ (handwritten)
$2^0 + 2^1 + 2^2 + \cdots + 2^{n-1}$ (handwritten)

It is then easy to verify that

$E[W_{n+1} \mid W_n = 1] = 1 = W_n$ (handwritten)

$\text{sum of } 2^n - 1$ (handwritten)

$$E(W_{n+1} \mid \mathcal{F}_n) = W_n,$$

$(2^n - 1) + 2^n$ (handwritten)

and hence W_n is a martingale with respect to $\mathcal{F}_n$.

$-(2^n - 1) - 2^n$ (handwritten)

Example 3. We can generalize the previous example. Suppose $X_1, X_2, \ldots$ are as in Example 2. Suppose that on the nth flip we make a bet equal to B_n. In determining the amount of the bet, we may look at the results of the first $(n-1)$st flips but cannot look beyond that. In other words, B_n is a random variable measurable with respect to $\mathcal{F}_{n-1}$ (we assume that B_1 is just a constant). The winnings after n flips, W_n, are given by $W_0 = 0$ and

$E(|B_n|) < \infty$ (handwritten)
no betting if an infinite amount (handwritten)

$$W_n = \sum_{j=1}^{n} B_j X_j.$$

$B_n \in \mathcal{F}_{n-1}$ (handwritten)

Assume that $E(|B_n|) < \infty$ (which will be guaranteed if the bet at time n is required to be less than some constant C_n). Then W_n is a martingale with respect to $\mathcal{F}_n$. To see this, we first note that $E(|W_n|) < \infty$ follows from the fact that $E(B_n) < \infty$ for each n. It is clear that W_n is $\mathcal{F}_n$ measurable. Finally,

analogous we see (handwritten)

$$E(W_{n+1} \mid \mathcal{F}_n) = E\left(\sum_{j=1}^{n+1} B_j X_j \mid \mathcal{F}_n\right)$$

let's define (handwritten)

$$= E\left(\sum_{j=1}^{n} B_j X_j \mid \mathcal{F}_n\right) + E(B_{n+1} X_{n+1} \mid \mathcal{F}_n).$$

W_n (handwritten)

I see the wage but not the outcome. (handwritten)

known today. (handwritten)

By (5.4),

$$E\left(\sum_{j=1}^{n} B_j X_j \mid \mathcal{F}_n\right) = \sum_{j=1}^{n} B_j X_j = W_n.$$

$E X_{n+1}$ so became we don't know (handwritten)

Since B_{n+1} is $\mathcal{F}_n$ measurable, it follows from (5.7) and (5.6) that

$$E(B_{n+1} X_{n+1} \mid \mathcal{F}_n) = B_{n+1} E(X_{n+1} \mid \mathcal{F}_n) = B_{n+1} E(X_{n+1}) = 0.$$

Therefore,

$$E(W_{n+1} \mid \mathcal{F}_n) = W_n.$$

Example 4. Polya's Urn. Consider an urn with balls of two colors, red and green. Assume that initially there is one ball of each color in the urn. At each time step, a ball is chosen at random from the urn. If a red ball is chosen, it is returned and in addition another red ball is added to the urn. Similarly, if a green ball is chosen, it is returned together with another green ball. Let X_n denote the number of red balls in the urn after n draws. Then $X_0 = 1$ and X_n is a (time-inhomogeneous) Markov chain with transitions

$$P\{X_{n+1} = k+1 \mid X_n = k\} = \frac{k}{n+2},$$

$$P\{X_{n+1} = k \mid X_n = k\} = \frac{n+2-k}{n+2}.$$

Let $M_n = X_n/(n+2)$ be the fraction of red balls after n draws. Then M_n is a martingale. To see this, note that

$$E(X_{n+1} \mid X_n) = X_n + \frac{X_n}{n+2}.$$

Since this is a Markov chain, all the relevant information in $\mathcal{F}_n$ for determining X_{n+1} is contained in X_n. Therefore,

$$
\begin{aligned}
E(M_{n+1} \mid \mathcal{F}_n) &= E((n+3)^{-1}X_{n+1} \mid X_n) \\
&= \frac{1}{n+3}[X_n + \frac{X_n}{n+2}] \\
&= \frac{X_n}{n+2} = M_n.
\end{aligned}
$$

A process M_n with $E(|M_n|) < \infty$ is called a *submartingale (supermartingale)* with respect to $X_0, X_1, \ldots$ if for each $m < n$, $E(M_n \mid \mathcal{F}_m) \geq (\leq) M_m$. In other words, a submartingale is a game in one's favor and a supermartingale is an unfair game. Note that M_n is a martingale if and only if it is both a submartingale and a supermartingale.

Example 5. Let X_n be a Markov chain with finite state space. Suppose a payoff function f is given as in Chapter 4. Let v be the value function associated to the payoff functions, $v(x) = E(f(X_T) \mid X_0 = x)$, where T is the stopping rule associated with the optimal strategy. Then $M_n = v(X_n)$ is a supermartingale with respect to $X_0, X_1, \ldots$.

5.3 Optional Sampling Theorem

The optional sampling theorem states in effect, "You cannot beat a fair game." However, it is easy to see that this theorem is false in complete generality. For example, suppose one plays the fair game of flipping a coin, winning one's bet if the coin comes up heads and losing one's bet if it is tails. Then using the "martingale betting strategy" described in Example 2 of Section 5.2, one can guarantee that one finishes the game ahead.

A stopping time T with respect to random variables $X_0, X_1, \ldots$ is a random variable taking values in the nonnegative integers (we allow $T = \infty$ as a possible value) that gives the time at which some procedure is stopped ($T = \infty$ corresponds to never stopping), such that the decision to stop must be made using only the information about $X_0, X_1, \ldots$ up to the present. Rigorously, we say that T is a stopping time (with respect to $X_0, \ldots, X_n$) if for each n, the indicator function of the event $\{T = n\}$ is measurable with respect to $X_0, \ldots, X_n$.

Example 1. Let k be an integer and let $T = k$.

Example 2. Let A be a set and let $T_A = \min\{j : X_j \in A\}$.

Example 3. If T and U are stopping times, then so are $\min\{T, U\}$ and $\max\{T, U\}$. In particular, if T is a stopping time and $T_n = \min\{T, n\}$, then each T_n is a stopping time, $T_0 \le T_1 \le T_2 \le \cdots$, and $T_n \le n$.

The optional sampling theorem states that (under certain conditions) if M_n is a martingale and T is a stopping time then $E(M_T) = E(M_0)$. This will not hold under all conditions since if we consider the martingale betting strategy and let T be the first time that the coin comes up heads, then $1 = E(M_T) \ne E(M_0) = 0$. The first thing we would like to show is that there is no way to beat a fair game if one has only a finite amount of time.

Fact. Suppose $M_0, M_1, \ldots$ is a martingale with respect to $X_0, X_1, \ldots$ and suppose T is a stopping time. Suppose that T is bounded, $T \le K$. Then

$$E(M_T \mid \mathcal{F}_0) = M_0.$$

In particular, $E(M_T) = E(M_0)$.

To prove this fact, we again write $\mathcal{F}_n$ for the information contained in $X_0, X_1, \ldots, X_n$. Note that $I\{T > n\}$, the indicator function of the event $\{T > n\}$, is measurable with respect to $\mathcal{F}_n$ (since we need only the information up through time n to determine if we have stopped by time n).

Since M_T is the random variable which equals M_j if $T = j$ we can write

$$M_T = \sum_{j=0}^{K} M_j I\{T = j\}.$$

Let us take the conditional expectation with respect to $\mathcal{F}_{K-1}$,

$$E(M_T \mid \mathcal{F}_{K-1}) = E(M_K I\{T = K\} \mid \mathcal{F}_{K-1})$$
$$+ \sum_{j=0}^{K-1} E(M_j I\{T = j\} \mid \mathcal{F}_{K-1}).$$

For $j \leq K - 1$, $M_j I\{T = j\}$ is $\mathcal{F}_{K-1}$ measurable; hence

$$E(M_j I\{T = j\} \mid \mathcal{F}_{K-1}) = M_j I\{T = j\}.$$

Since T is known to be no more than K, the event $\{T = K\}$ is the same as the event $\{T > K - 1\}$. The latter event is measurable with respect to $\mathcal{F}_{K-1}$. Hence, using (5.7),

$$\begin{aligned} E(M_K I\{T = K\} \mid \mathcal{F}_{K-1}) &= E(M_K I\{T > K - 1\} \mid \mathcal{F}_{K-1}) \\ &= I\{T > K - 1\} E(M_K \mid \mathcal{F}_{K-1}) \\ &= I\{T > K - 1\} M_{K-1}. \end{aligned}$$

The last equality follows from the fact the M_n is a martingale. Therefore,

$$\begin{aligned} E(M_T \mid \mathcal{F}_{K-1}) &= I\{T > K - 1\} M_{K-1} + \sum_{j=0}^{K-1} M_j I\{T = j\} \\ &= I\{T > K - 2\} M_{K-1} + \sum_{j=0}^{K-2} M_j I\{T = j\}. \end{aligned}$$

If we work through this argument again, this time conditioning with respect to $\mathcal{F}_{K-2}$, we get

$$\begin{aligned} E(M_T \mid \mathcal{F}_{K-2}) &= E(E(M_T \mid \mathcal{F}_{K-1}) \mid \mathcal{F}_{K-2}) \\ &= I\{T > K - 3\} M_{K-2} + \sum_{j=0}^{K-3} M_j I\{T = j\}. \end{aligned}$$

We can continue this process until we get

$$E(M_T \mid \mathcal{F}_0) = M_0.$$

There are many examples of interest where the stopping time T is not bounded. Suppose T is a stopping time with $P\{T < \infty\} = 1$, i.e., a rule that guarantees that one stops eventually (The time associated to the martingale betting strategy satisfies this condition). When can we conclude that $E(M_T) = E(M_0)$? To investigate this consider the stopping times $T_n = \min\{T, n\}$. Note that

$$M_T = M_{T_n} + M_T I\{T > n\} - M_n I\{T > n\}.$$

Hence,

$$E(M_T) = E(M_{T_n}) + E(M_T I\{T > n\}) - E(M_n I\{T > n\}).$$

Since T_n is a bounded stopping time, it follows from the above that $E(M_{T_n}) = E(M_0)$. We would like to be able to say that the other two terms do not contribute as $n \to \infty$. The second term is not much of a problem. Since the probability of the event $\{T > n\}$ goes to 0 as $n \to \infty$, we are taking the expectation of the random variable M_T restricted to a smaller and smaller set. One can show (see Section 5.4) that if $E(|M_T|) < \infty$ then $E(|M_T|I\{T > n\}) \to 0$.

The third term is more troublesome. Consider Example 2 of Section 5.2 again. In this example, the event $\{T > n\}$ is the event that the first n flips are tails and has probability 2^{-n}. If this event occurs, the bettor has lost a total of $2^n - 1$ dollars, i.e., $M_n = 1 - 2^n$. Hence

$$E(M_n I\{T > n\}) = 2^{-n}(1 - 2^n),$$

which does not go to 0 as $n \to \infty$. This is why the desired result fails in this case. However, if M_n and T are given so that

$$\lim_{n \to \infty} E(|M_n|I\{T > n\}) = 0,$$

then we will be able to conclude that $E(M_T) = E(M_0)$. We summarize this as follows.

Optional Sampling Theorem. Suppose $M_0, M_1, \ldots$ is a martingale with respect to $X_0, X_1, \ldots$ and T is a stopping time satisfying $P\{T < \infty\} = 1$,

$$E(|M_T|) < \infty, \tag{5.10}$$

and

$$\lim_{n \to \infty} E(|M_n|I\{T > n\}) = 0. \tag{5.11}$$

Then,

$$E(M_T) = E(M_0).$$

Example 1. Let X_n be a simple random walk $(p = 1/2)$ on $\{0, \ldots, N\}$ with absorbing boundaries. Suppose $X_0 = a$. Then, X_n is a martingale. Let $T = \min\{j : X_j = 0 \text{ or } N\}$. T is a stopping time, and since X_n is bounded, (5.10) and (5.11) are satisfied [note that (5.10) and (5.11) are always satisfied if the martingale is bounded and $P\{T < \infty\} = 1$]. Therefore

$$E(M_T) = E(M_0) = a.$$

But in this case $E(M_T) = NP\{X_T = N\}$. Therefore,

$$P\{X_T = N\} = a/N.$$

This gives another derivation of the gambler's ruin result for simple random walk.

Example 2. Let X_n be as in Example 1 and let $M_n = X_n^2 - n$. Then M_n is a martingale with respect to X_n. To see this, note that by Example 2, Section 5.1,

$$E(M_{n+1} \mid \mathcal{F}_n) = E(X_{n+1}^2 - (n+1) \mid \mathcal{F}_n\} = X_n^2 + 1 - (n+1) = M_n.$$

Again, let $T = \min\{j : X_j = 0 \text{ or } N\}$. In this case, M_n is not a bounded martingale so it is not immediate that (5.10) and (5.11) hold. However, one can prove (Exercise 1.7) that there exist $C < \infty$ and $\rho < 1$ such that

$$P\{T > n\} \le C\rho^n.$$

Since $|M_n| \le N^2 + n$, one can then show that $E(|M_T|) < \infty$ and

$$E(|M_n|I\{T > n\}) \le C\rho^n(N^2 + n) \to 0.$$

Hence the conditions of the optional sampling theorem hold and we can conclude

$$E(M_T) = E(M_0) = a^2.$$

Note that

$$E(M_T) = E(X_T^2) - E(T) = N^2P\{X_T = N\} - E(T) = aN - E(T).$$

Hence,

$$E(T) = aN - a^2 = a(N - a).$$

Example 3. Let X_n be a simple random walk $(p = 1/2)$ on the integers $\{\ldots, -1, 0, 1, \ldots\}$ with $X_0 = 0$. We have seen that this is a martingale. Let $T = \min\{j : X_j = 1\}$. Since simple random walk is recurrent, $P\{T < \infty\} = 1$. Note that $X_T = 1$ and hence

$$1 = E(X_T) \ne E(X_0) = 0.$$

Therefore, the conditions of the optional sampling theorem must not hold. We will not give the details here but it can be shown in this case that $P\{T > n\} \sim cn^{-1/2}$ for some constant c. By the central limit theorem, the random walk tends to go a distance of order $\sqrt{n}$ in n steps. In this case $E(|X_n|I\{T > n\})$ does not go to 0.

5.4 Uniform Integrability

Condition (5.11) is often hard to verify. For this reason we would like to describe some conditions that imply (5.11) that may be easier to verify.

We start by considering one random variable X with $E(|X|) < \infty$. Let F denote the distribution function for $|X|$. Then it follows that

$$\lim_{n \to \infty} E(|X|I\{|X| > n\}) = \lim_{n \to \infty} \int_n^\infty |x|\, dF(x) = 0.$$

Suppose $P\{|X| > n\} = \delta$ and suppose A is any other event with $P(A) = \delta$. Since $\{|X| > n\}$ is the event on which $|X|$ is the largest, it is easy to see that

$$E(|X|I_A) \le E(|X|I\{|X| > n\}).$$

From this we can see the following: if X is a random variable with $E(|X|) < \infty$ then for every $\epsilon > 0$ there exists a $\delta > 0$ such that if $P(A) < \delta$ then $E(|X|I_A) < \epsilon$.

Now suppose we have a sequence of random variables $X_1, X_2, \ldots$. We say that the sequence is *uniformly integrable* if for every $\epsilon > 0$ there exists a $\delta > 0$ such if $P(A) < \delta$, then

$$E(|X_n|I_A) < \epsilon \tag{5.12}$$

for each n. The key parts of the definition are that δ may not depend on n and (5.12) must hold for all values of n.

To develop a sense of the definition, we will start by giving an example of random variables that are not uniformly integrable. Consider Example 2 of Section 5.2, the martingale betting strategy, and consider the random variables $W_0, W_1, W_2, \ldots$. If we let A_n be the event $\{X_1 = X_2 = \cdots = X_n = -1\}$ then $P(A_n) = 2^{-n}$ and $E(|W_n|I_{A_n}) = 2^{-n}(2^n - 1) \to 1$. We clearly cannot satisfy the conditions for uniform integrability for any $\epsilon < 1$.

Now suppose that $M_0, M_1, \ldots$ is a uniformly integrable martingale with respect to $X_0, X_1, \ldots$ and T is a stopping time with $P\{T < \infty\} = 1$. Then

$$\lim_{n \to \infty} P\{T > n\} = 0,$$

and hence by uniformly integrability

$$\lim_{n \to \infty} E(|M_n|I\{T > n\}) = 0;$$

that is, (5.11) holds. We can therefore give another statement of the optional sampling theorem.

Optional Sampling Theorem. Suppose $M_0, M_1, \ldots$ is a uniformly integrable martingale with respect to $X_0, X_1, \ldots$ and T is a stopping time satisfying $P\{T < \infty\} = 1$ and $E(|M_T|) < \infty$. Then $E(M_T) = E(M_0)$.

The condition of uniform integrability can be difficult to verify. There are a number of easier conditions that imply uniform integrability. We mention one here and give another in the exercises (Exercise 5.13).

Fact. If $X_1, X_2, \ldots$ is a sequence of random variables and there exists a

$C < \infty$ such that $E(X_n^2) < C$ for each n then the sequence is uniformly integrable.

To prove the fact, let $\epsilon > 0$ be given and let $\delta = \epsilon^2/4C$. Suppose $P(A) < \delta$. Then

$$
\begin{aligned}
E(|X_n|I_A) &= E[|X_n|I(A \cap \{|X_n| \geq 2C/\epsilon\})] \\
&\quad + E[|X_n|I(A \cap \{|X_n| < 2C/\epsilon\})] \\
&\leq (\epsilon/2C)E[|X_n|^2 I(A \cap \{|X_n| \geq 2C/\epsilon\})] \\
&\quad + (2C/\epsilon)P(A \cap \{|X_n| < 2C/\epsilon\} \\
&\leq (\epsilon/2C)E(|X_n|^2) + (2C/\epsilon)P(A) < \epsilon.
\end{aligned}
$$

Example 1. Random Harmonic Series. It is well known that the harmonic series $1 + \frac{1}{2} + \frac{1}{3} + \cdots$ diverges while the alternating harmonic series $1 - \frac{1}{2} + \frac{1}{3} - \frac{1}{4} + \cdots$ converges. What if the pluses and minuses are chosen at random? To model this, let $X_1, X_2, \ldots$ be independent random variables with $P\{X_i = 1\} = P\{X_i = -1\} = 1/2$. Let $M_0 = 0$ and for $n > 0$,

$$
M_n = \sum_{j=1}^{n} \frac{1}{j} X_j.
$$

By Example 1, Section 5.2, M_n is a martingale. Since $E(M_n) = 0$,

$$
E(M_n^2) = Var(M_n^2) = \sum_{j=1}^{n} Var\left(\frac{1}{j} X_j\right) = \sum_{j=1}^{n} \frac{1}{j^2} \leq \sum_{j=1}^{\infty} \frac{1}{j^2} < \infty.
$$

Hence M_n is a uniformly integrable martingale. The question of convergence is discussed in the next section.

Example 2. Branching Process. Let X_n denote the number of offspring in the nth generation of a branching process (see Chapter 2, Section 2.4) whose offspring distribution has mean μ and variance σ^2. Then (Exercise 5.4) $M_n = \mu^{-n} X_n$ is a martingale with respect to $X_1, X_2, \ldots$. Suppose $\mu > 1$. Then (Exercise 5.8) there exists a constant such that for all n, $E(M_n^2) < \infty$ and hence M_n is a uniformly integrable martingale for $\mu > 1$.

5.5 Martingale Convergence Theorem

The martingale convergence theorem states that under very general conditions a martingale M_n converges to a limiting random variable M_∞. We start by considering a particular example, Polya's urn (Example 4, Section 5.2). In this case M_n is the proportion of red balls in the urn after n draws. What happens as n gets large? In Exercise 5.9 it is shown that the distribution of M_n is approximately a uniform distribution on $[0, 1]$ for large values of n. This leads to a question: Does the proportion of red balls fluctuate between 0 and 1 infinitely often or does it eventually settle down to a particular value? We will show now that the latter is true.

Let $0 < a < b < \infty$ and suppose that $M_n < a$. Let T be the stopping time

$$T = \min\{j : j \geq n \text{ and } M_j \geq b\},$$

and let $T_m = \min\{T, m\}$. Then for $m > n$, the optional sampling theorem states that

$$E(M_{T_m}) = M_n < a.$$

But

$$E(M_{T_m}) \geq E(M_{T_m} I\{T \leq m\}) = E(M_T I\{T \leq m\}) \geq b P\{T \leq m\}.$$

Hence,

$$P\{T \leq m\} < a/b.$$

Since this is true for all m,

$$P\{T < \infty\} \leq a/b.$$

Doubling Strategy

$M_n = \{X_1, X_2, X_3 \cdots X_n = 1\}$

n straight loss

$P(A_n) \geq 2^n$ $E\left(|M_n| \cdot \mathbb{I}_{A_n}\right) = (2^n - 1) \cdot 2^{-n}$

This says that with probability of at least $1 - (a/b)$ the proportion of red balls never gets as high as b. Now suppose the proportion of red balls does get higher than b. What then is the probability that the proportion goes down below a again? By the same argument applied to the proportion of green balls we can say that the probability of dropping below a is at most $(1 - b)/(1 - a)$. By continuing this argument, we can see that, starting at a, the probability of going above b, then below a again, then above b, then below a, a total of n times, can be bounded above by

$$\left(\frac{a}{b}\right)\left(\frac{1-b}{1-a}\right)\left(\frac{a}{b}\right)\cdots\left(\frac{a}{b}\right)\left(\frac{1-b}{1-a}\right) = \left(\frac{a}{b}\right)^n\left(\frac{1-b}{1-a}\right)^n,$$

which tends to 0 as $n \to \infty$. Hence, the proportion does not fluctuate infinitely often between a and b. Since a and b are arbitrary, this shows that it is impossible for the proportion to fluctuate infinitely often between any two numbers, or, in other words, the limit

$$M_\infty = \lim_{n \to \infty} M_n$$

exists. The limit M_∞ is a random variable; it is not difficult to show (see Exercise 5.9) that M_∞ has a uniform distribution on $[0, 1]$.

We now state a general result.

Martingale Convergence Theorem. Suppose $M_0, M_1, \ldots$ is a martingale with respect to $X_0, X_1, \ldots$ such that there exists a $C < \infty$ with $E(|M_n|) < C$ for all n. Then there exists a random variable M_∞ such that

$$M_n \to M_\infty.$$

The proof of this theorem is similar to the argument above. What we will show is that for every $0 < a < b < \infty$ the probability that the martingale fluctuates infinitely often between a and b is 0. Since this will be true for every $a < b$, it must be the case that the martingale M_n converges to some value M_∞.

Fix $a < b$. We will consider the following betting strategy which is reminiscent of the martingale betting strategy. We think of M_n as giving the cumulative results of some fair game and $M_{n+1} - M_n$ as being the result of the game at time $n + 1$. Whenever $M_n < a$, bet 1 on the martingale. Continue this procedure until the martingale gets above b. Then stop betting until the martingale drops below a again and return to betting 1. Continue this process, changing the bet to 0 when M_n goes above b and changing back to 1 when M_n drops below a. Note that if the martingale fluctuates infinitely often between a and b this gives a winning strategy.

After n steps the winnings in this strategy are given by

$$W_n = \sum_{j=1}^{n} B_j(M_j - M_{j-1}),$$

where B_j is the bet which equals 1 or 0 depending on whether the martingale was most recently below a or above b. One can verify as in Example 3, Section 5.2 that W_n is a martingale with respect to $M_0, M_1, \ldots$. We note that

$$W_n \geq (b - a)U_n - |M_n - a|,$$

where U_n denotes the number of times that the martingale goes between a and b (this is often called the number of "upcrossings") and $|M_n - a|$ gives an estimate for the amount lost in the last interval (this is relevant if the bettor is betting 1 at time n). Since W_n is a martingale we have

$$E(W_0) = E(W_n) \geq (b - a)E(U_n) - E(|M_n - a|).$$

Since $E(|M_n - a|) \leq E(|M_n|) + a \leq C + a$, we get

$$E(U_n) \leq (b - a)^{-1}[E(W_0) + C + a].$$

Since this holds for every n, the expected number of upcrossings up to infinity is bounded and hence with probability 1 the number of upcrossings is finite. This proves the theorem.

It follows immediately from the martingale property that for every n, $E(M_n) = E(M_0)$. It is not necessarily true, however, that $E(M_\infty) = E(M_0)$. For a counterexample, we return to the martingale betting strategy. In this case

$$W_\infty = \lim_{n \to \infty} W_n = 1,$$

and hence $E(W_\infty) \neq E(W_0) = 0$. If the martingale is uniformly integrable, it is true that the limiting random variable has the same expectation (see Exercise 5.11).

Fact. If M_n is a uniformly integrable martingale with respect to X_0, X_1, ..., then

$$M_\infty = \lim_{n \to \infty} M_n,$$

exists and $E(M_\infty) = E(M_0)$.

Example 1. Let X_n be the number of individuals in the nth generation of a branching process whose offspring distribution has mean μ and variance σ^2. Assume $X_0 = 1$ and let $M_n = \mu^{-n} X_n$ be the associated martingale. If $\mu \leq 1$, we know that extinction occurs with probability 1 and hence $M_n \to M_\infty = 0$. In this case $E(M_\infty) \neq E(M_0)$. In Section 5.4, we noted that M_n is uniformly integrable if $\mu > 1$, and hence M_∞ is a nontrivial random variable with $E(M_\infty) = 1$.

Example 2. Let $X_1, X_2, \ldots$ be independent random variables with $P\{X_i = 1\} = P\{X_i = -1\} = 1/2$ and let

$$M_n = \sum_{j=1}^{n} \frac{1}{j} X_j.$$

It was noted in Section 5.4 that M_n is a uniformly integrable martingale. Hence M_n approaches a random variable M_∞. This says that the random harmonic series converges with probability 1.

Example 3. Let M_n be the proportion of red balls in Polya's urn. In this case, suppose that at time $n = 0$ there are k red balls and m green balls (so after n draws there are $n + k + m$ balls). Since M_n is bounded it is easy to see that M_n is a uniformly integrable martingale and M_n approaches a random variable M_∞ with $E(M_\infty) = E(M_0) = k/(k+m)$. It can be shown that the distribution of M_∞ is a beta distribution with parameters k and m, i.e., it has density

$$\frac{\Gamma(k+m+2)}{\Gamma(k)\Gamma(m)} x^{k-1}(1-x)^{m-1}, \quad 0 < x < 1.$$

This distribution arises naturally in Bayesian statistics. Suppose one is interested in estimating the probability p that some event occurs. If one has no information about p then one might initially say that p is a random variable with a uniform distribution on $[0, 1]$. Now suppose that one performs the trial $k + m - 2$ times and the event occurs $k - 1$ times. Then one's posterior distribution (using Bayes theorem) for the probability p is given by the beta distribution with parameters k and m. It can easily be verified that the expected value of a beta random variable is $k/(k+m)$.

Example 4. Let M_n be a martingale with respect to $X_0, X_1, \ldots$, and let T be a stopping time with $P\{T < \infty\} = 1$. Let $T_n = \min\{n, T\}$ and $Y_n = M_{T_n}$. Then $Y_n \to Y_\infty$ where $Y_\infty = M_T$. As we saw in the optional sampling theorem, it is not always the case that $E(Y_\infty) = E(Y_0)$. However, this is true if M_n is uniformly integrable.

Example 5. Let X_n be an irreducible Markov chain on a countably infinite state space S with transition function $p(x, y)$. A function f is called harmonic at x if

$$f(x) = \sum_{y \in S} p(x, y) f(y).$$

In Chapter 2 we considered the problem of determining whether or not the chain was recurrent. We now prove one of the assertions we made there. Suppose z is a fixed state in S and let $u(x)$ denote the probability starting at state x that the chain ever reaches state z. In other words, if

$$T = \min\{j \geq 0 : X_j = z\},$$

then

$$u(x) = P\{T < \infty \mid X_0 = x\}.$$

As we noted then, $u(z) = 1$ and $u(x)$ is harmonic at any $x \neq z$.

Suppose now that we can find some function v that satisfies:

$$v(z) = 1, \tag{5.13}$$

$$0 \leq v(x) \leq 1, \tag{5.14}$$

$$v(x) = \sum_{y \in S} p(x, y) v(y), \quad x \neq z. \tag{5.15}$$

If T is defined as above, and $T_n = \min\{n, T\}$, one can check that $M_n = v(X_{T_n})$ is a martingale with respect to $X_0, X_1, \ldots$. Since v is bounded, M_n is uniformly integrable and

$$\lim_{n \to \infty} M_n = M_\infty,$$

exists with $E(M_\infty) = E(M_0)$.

If the chain is recurrent, then $P\{T < \infty\} = 1$ and $M_\infty = v(z) = 1$. Hence if $X_0 = x$, $1 = E(M_0) = v(x)$. Thus, if the chain is recurrent there is no nontrivial solution to equations (5.13) – (5.15).

Example 6. Let $X_1, X_2, \ldots$ be independent random variables with

$$P\{X_i = 3/2\} = P\{X_i = 1/2\} = 1/2.$$

Let $M_0 = 1$ and for $n > 0$, let $M_n = X_1 \cdots X_n$. Note that $E(M_n) = E(X_1) \cdots E(X_n) = 1$, and in fact, if $\mathcal{F}_n$ denotes the information contained in $X_1, \ldots, X_n$,

$$
\begin{aligned}
E(M_{n+1} \mid \mathcal{F}_n) &= E(X_1 \cdots X_{n+1} \mid \mathcal{F}_n) \\
&= X_1 \cdots X_n E(X_{n+1} \mid \mathcal{F}_n) \\
&= X_1 \cdots X_n E(X_{n+1}) = M_n.
\end{aligned}
$$

Hence M_n is a martingale with respect to $X_1, X_2, \ldots$. Since $E(|M_n|) = E(M_n) = 1$, the conditions of the martingale convergence theorem hold and hence

$$M_n \to M_\infty$$

for some random variable M_∞. Is M_n uniformly integrable? The answer is no; in fact, the limiting random variable $M_\infty = 0$ [and hence $E(M_\infty) \neq E(M_0)$]. To see this, consider the logarithm of the martingale,

$$\ln M_n = \sum_{j=1}^n \ln X_j.$$

The right-hand side is the sum of independent identically distributed random variables with mean

$$E(\ln X_i) = \frac{1}{2} \ln \frac{1}{2} + \frac{1}{2} \ln \frac{3}{2} < 0.$$

By the law of large numbers, $\ln M_n \to -\infty$ and hence $M_n \to 0$.

Note in this case

$$E(M_n^2) = E(X_1^2) \cdots E(X_n^2) = (5/4)^n,$$

so the second moment is not uniformly bounded.

5.6 Exercises

5.1 Consider the experiment of rolling two dice. Let X be the value of the first roll and Y the sum of the two dice. Find $E(X \mid Y)$, i.e., give the value of $E(X \mid Y)(y)$ for all y.

5.2 Suppose that X_t is a Poisson process with parameter $\lambda = 1$. Find $E(X_1 \mid X_2)$ and $E(X_2 \mid X_1)$.

5.3 Let $X_1, X_2, X_3, \ldots$ be independent identically distributed random variables. Let $m(t) = E(e^{tX_1})$ be the moment generating function of X_1 (and hence of each X_i). Fix t and assume $m(t) < \infty$. Let $S_0 = 0$ and for $n > 0$,

$$S_n = X_1 + \cdots + X_n.$$

Let $M_n = m(t)^{-n} e^{tS_n}$. Show that M_n is a martingale with respect to $X_1, X_2, \ldots$.

5.4 Let $X_0, X_1, \ldots$ be the values of a branching process as in Chapter 2, Section 2.4, i.e., X_n gives the number of individuals in the nth generation. Suppose that the mean number of offspring per individual is μ. Show that $M_n = \mu^{-n} X_n$ is a martingale with respect to $X_0, X_1, \ldots$

5.5 COMPUTER SIMULATION. Consider the Polya urn model. Simulate this model with a computer by starting with one red and one green ball and continuing until the number of balls in the urn is 1000. Note the fraction of red balls in the 1000 balls. Do this simulation at least 2000 times and note how many times the fraction of red balls is in the intervals $[0, .05), [.05, .1), \ldots, [.95, 1)$. From the simulation data, make a conjecture as to what the distribution of the fraction of red balls looks like.

5.6 Consider a biased random walk on the integers with probability $p < 1/2$ of moving to the right and probability $1 - p$ of moving to the left. Let S_n be the value at time n and assume that $S_0 = a$, where $0 < a < N$.
  (a) Show that $M_n = [(1 - p)/p]^{S_n}$ is a martingale.
  (b) Let T be the first time that the random walk reaches 0 or N, i.e.,

$$T = \min\{n : S_n = 0 \text{ or } N\}.$$

Use optional sampling on the martingale M_n to compute $P\{S(T) = 0\}$.

5.7 Let S_n be as in Exercise 5.6.
  (a) Show that $M_n = S_n + (1 - 2p)n$ is a martingale.
  (b) Let T be the first time that the random walk reaches 0 or N, i.e.,

$$T = \min\{n : S_n = 0 \text{ or } N\}.$$

Let $T_n = \min\{n, T\}$ and let Z_n be the martingale $Z_n = M_{T_n}$. Show that there exists a $C < \infty$ such that $E(Z_n^2) < C$ for all n. You may wish to use Exercise 1.7.
  (c) Apply the optional sampling theorem to $E(M_T)$ and use this and the result from Exercise 5.6 to find the expected number of steps until absorption, $E(T)$.

5.8 Let X_n be the number of individuals in the nth generation of a branching process in which each individual produces offspring from a distribution with mean μ and variance σ^2. We have seen previously that $M_n = \mu^{-n} X_n$ is a martingale.
  (a) Let $\mathcal{F}_n$ denote the information contained in $X_0, \ldots, X_n$. Show that

$$E(X_{n+1}^2 \mid \mathcal{F}_n) = \mu^2 X_n^2 + \sigma^2 X_n.$$

  (b) Suppose $\mu > 1$. Show that there exists a $C < \infty$ such that for all n

$$E(M_n^2) < C.$$

  (c) Show that this is not the case if $\mu \leq 1$.

5.9 Consider the Polya urn problem. Let M_n be the proportion of red balls after n draws (starting with one red and one green ball). Prove by induction on n that

$$P\{M_n = \frac{k}{n + 2}\} = \frac{1}{n + 1}, \quad k = 1, 2, \ldots, n + 1.$$

5.10 Do another simulation of the Polya urn model. Again, start with one red and one green ball and continue until there are 1000 balls in the urn. Note the proportion of red balls at this time and then continue until there are 2000 balls. Compare these two numbers (i.e., compare M_{998} and M_{1998}). Do this at least 30 times.

5.11 Suppose $X_1, X_2, \ldots$ are uniformly integrable with $X_n \to Y$ with probability 1. Show that $E(X_n) \to E(Y)$.

5.12 Let $X_1, X_2, \ldots$ be independent, identically distributed random variables taking values in $\{-1, 0, 1, \ldots\}$ with mean $\mu < 0$. Let $S_0 = 1$ and for $n > 0$,

$$S_n = 1 + S_1 + \ldots + S_n.$$

Let $T = \min\{n : S_n = 0\}$. By the law of large numbers, we know that $P\{T < \infty\} = 1$. Show that $E(T) \leq 1/|\mu|$. [Hint: it suffices to prove for each n, if $T_n = \min\{n, T\}$, then $E(T_n) \leq 1/|\mu|$. Consider the martingale $M_n = S_n - n\mu$.] Exercise 5.14 below can be used to prove that $E(T) = 1/|\mu|$.

5.13 Let M_n be a martingale with respect to $\mathcal{F}_n$. Assume there exists a nonnegative random variable Y with $E(Y) < \infty$ and $|M_n| \leq Y$ for all n. Show that M_n is a uniformly integrable martingale.

5.14 Let $X_1, X_2, \ldots$ be independent, identically distributed random variables with mean μ. Let T be a stopping time with respect to $X_1, X_2, \ldots$ with $E(T) < \infty$.
  (a) Let

$$Y = \sum_{n=1}^{\infty} |X_n| I\{T \geq n\},$$

where I denotes the indicator function. Show that $E(Y) < \infty$.
  (b) Let $T_n = \min\{n, T\}$ and

$$M_n = X_1 + \cdots + X_{T_n} - \mu T_n.$$

Explain why M_n is a uniformly integrable martingale (see Exercise 5.13).
  (c) Prove Wald's equation,

$$E\left(\sum_{n=1}^{T} X_n\right) = \mu E(T).$$

CHAPTER 6

Renewal Processes

6.1 Introduction

Let $T_1, T_2, \ldots$ be independent, identically distributed, nonnegative random variables with distribution function $F(x) = P\{T_i \leq x\}$. We can think of the random variables T_i as being the lifetimes of a component or as the times between occurrences of some event. The renewal process associated with T_i is the process N_t with $N_t = 0$ for $t < T_1$ and otherwise

$$N_t = \max\{n : T_1 + \cdots + T_n \leq t\}.$$

N_t denotes the number of events that have occurred up to time t. Sometimes we will consider a slightly more general process where Y is a nonnegative random variable independent of $T_1, T_2, \ldots$, with perhaps a different distribution. We think of Y as the time until the first event, and then the waiting times for later events are given by the T_i. More precisely, we set $N_t = 0$ for $t < Y$; and for $t \geq Y$,

$$N_t = \min\{n : Y + T_1 + \cdots + T_n > t\}. \tag{6.1}$$

We will assume that the random variables T_i have finite, positive mean and we set

$$\mu = E(T_i).$$

Example 1. Poisson Process. Consider the Poisson process with rate parameter λ. The waiting times $T_1, T_2, \ldots$ are independent, exponential random variables with parameter λ and N_t is the Poisson process. In this case $\mu = 1/\lambda$.

Example 2. Let X_n be an irreducible, positive recurrent, discrete-time Markov chain starting in state x. Let

$$T_1 = \min\{n > 0 : X_n = x\},$$

and for $i > 1$ let

$$T_i = \min\{n > 0 : X_{T_1 + \cdots + T_{i-1} + n} = x\}.$$

In other words, T_i measures the amount of time between the $(i-1)$st return and the ith return to state x . In general it is difficult to determine the distribution function F for T_i given the transition matrix for the chain. We noted previously [see (1.11)] that

$$E(T_i) = \frac{1}{\pi(x)},$$

where π denotes the invariant probability measure for the chain. If we instead start the chain at some state $y \neq x$ we can define

$$Y = \min\{n > 0 : X_n = x\},$$

$$T_1 = \min\{n > 0 : X_{Y+n} = x\},$$

and recursively,

$$T_i = \min\{n > 0 : X_{Y+T_1+\cdots+T_{i-1}+n} = x\}.$$

Example 3. Let X_t be an irreducible, positive recurrent, continuous-time Markov chain starting in state x. Define

$$R_1 = \inf\{t > 0 : X_t \neq x\},$$

$$S_1 = \inf\{t > R_1 : X_t = x\},$$

$$T_1 = R_1 + S_1,$$

and in general

$$R_i = \inf\{t > 0 : X_{T_1+\cdots+T_{i-1}+t} \neq x\},$$

$$S_i = \inf\{t > 0 : X_{T_1+\cdots+T_{i-1}+R_i+t} = x\},$$

$$T_i = R_i + S_i.$$

The random variables R_i are exponential with parameter $\alpha(x)$, the rate at which the chain is changing from state x. The distribution of the S_i, and hence the T_i, is not so easy to determine.

Example 4. $M/G/1$ **Queue.** We will imagine that we have a queue with a single server. Customers arrive according to a Poisson process with rate λ, i.e., the waiting times between customer arrivals are independent exponential random variables with parameter λ. We will assume that the service times for customers are independent, identically distributed random variables with mean μ. However, we will not assume that the service times are exponential (in most cases of interest one does not expect that the service time should have the "loss of memory" property so an exponential distribution is not appropriate). The G in $M/G/1$ stands for "general" (service distribution).

If we let Y_t denote the number of people in the queue, then Y_t is not a Markov process. However there is a natural renewal process one can associate with the queue. Suppose $Y_0 = 0$. Let

$$R_1 = \inf\{t > 0 : R_1 = 1\},$$
$$S_1 = \inf\{t > 0 : Y_{R_1+t} = 0\},$$
$$T_1 = R_1 + S_1.$$

Similarly, we define for $i > 1$,

$$R_i = \inf\{t > 0 : Y_{T_1+\cdots+T_{i-1}+t} = 1\},$$
$$S_i = \inf\{t > 0 : Y_{T_1+\cdots+T_{i-1}+R_i+t} = 0\},$$
$$T_i = R_i + S_i.$$

Note that the variables R_i are exponential with rate λ, but the distribution of the S_i can be very complicated. Nevertheless, under the assumption that $E(T_i) < \infty$, we can see that $T_1, T_2, \cdots$ satisfy the conditions for a renewal process. We can think of the time represented by the R_i as the "idle times" and the time represented by the S_i as the "busy times."

Suppose we have a renewal process N_t corresponding to the random variables $T_1, T_2, \ldots$. In general, N_t is not a Markov process; in order to predict when the next occurrence will happen we need to know when the last occurrence took place. For this reason it is natural to consider the "age process"

$$A_t = \begin{cases} t, & \text{if } N_t = 0, \\ t - [T_1 + \cdots + T_{N_t}], & \text{if } N_t > 0. \end{cases}$$

The process (N_t, A_t) can be thought of as a Markov process. The Poisson process is a special example of a renewal process that *is* a Markov process; for the Poisson process the probability of an event occurring in the interval $[t, t + \Delta t]$ is independent of A_t. This follows from the "loss of memory" property associated with the exponential distribution.

The first result for renewal processes will be the analogue of the (strong) law of large numbers. Recall that the law of large numbers states that with probability 1,

$$\lim_{n\to\infty} \frac{T_1 + \cdots + T_n}{n} = \mu.$$

In terms of the renewal process N_t this states that for all $\epsilon > 0$, if n is sufficiently large,

$$N_{\mu n(1-\epsilon)} \leq n;$$
$$N_{\mu n(1+\epsilon)} \geq n.$$

Equivalently, for all $\epsilon > 0$, it t is sufficiently large,

$$N_t \leq \frac{t}{\mu(1-\epsilon)},$$
$$N_t \geq \frac{t}{\mu(1+\epsilon)}.$$

In other words, with probability 1,

$$\lim_{t \to \infty} \frac{N_t}{t} = \frac{1}{\mu}. \tag{6.2}$$

We now derive a central limit theorem for renewal processes. Assume that the variance of each T_i is $\sigma^2 < \infty$. Then the usual central limit theorem states that the distribution of

$$\frac{T_1 + \cdots + T_n - n\mu}{\sigma \sqrt{n}}$$

approaches a unit normal (i.e., a normal random variable with mean 0, variance 1). Slightly more informally we can say that for large n

$$T_1 + \cdots + T_n \approx n\mu + \sigma \sqrt{n} B,$$

where B is a unit normal. This states that the number of occurrences in time $n\mu + \sigma \sqrt{n} B$ is n. From (6.2), we would expect the number of occurrences in the time interval of size $\sigma \sqrt{n} |B|$ to be about $\sigma \sqrt{n} |B|/\mu$. Hence we have the number of occurrences in time $n\mu$ is about

$$n - \frac{\sigma}{\mu} \sqrt{n} B.$$

If we write t for $n\mu$ and note that $-B$ is also a unit normal random variable we see that

$$N_t \approx \frac{t}{\mu} + \frac{\sigma}{\mu^{3/2}} \sqrt{t} B,$$

where B is a unit normal. While this is only a rough sketch, this argument can be made rigorous, giving a central limit theorem for renewal processes.

Theorem. If the waiting times T_i have mean μ and variance σ^2, then as $t \to \infty$ the distribution of

$$\frac{N_t - \mu^{-1} t}{\sigma \mu^{3/2} \sqrt{t}}$$

approaches a standard normal distribution.

Example 5. We will apply the above informal reasoning to a somewhat more complicated example. Suppose we have a continuous-time Markov chain X_t on state space $\{1, 2\}$ with $\alpha(1, 2) = \alpha_1$ and $\alpha(2, 1) = \alpha_2$. Assume $X_0 = 1$ and let Y_t denote the amount of time spent in state 1 up to time t,

$$Y_t = \int_0^t I\{X_s = 1\} \, ds.$$

Define R_i and S_i as in Example 3 above (with $x = 1$). The random variables R_i are exponential with rate α_1 and hence have mean $\mu_1 = 1/\alpha_1$ and variance $\sigma_1^2 = 1/\alpha_1^2$. Similarly the random variables S_i are exponential with mean $\mu_2 = 1/\alpha_2$ and variance $\sigma_2^2 = 1/\alpha_2^2$. For large n the central limit theorem states that

$$R_1 + \cdots + R_n \approx n\mu_1 + \sigma_1\sqrt{n}B_1,$$

$$S_1 + \cdots + S_n \approx n\mu_2 + \sigma_2\sqrt{n}B_2,$$

where B_1 and B_2 are independent unit normals. In other words, in time $n(\mu_1 + \mu_2) + \sqrt{n}(\sigma_1 B_1 + \sigma_2 B_2)$, the amount of time spent in state 1 is approximately $n\mu_1 + \sqrt{n}\sigma_1 B_1$. For large t, the amount of time spent in state 1 in an interval $[t, t+\Delta t]$ is about $\Delta t[\mu_1/(\mu_1+\mu_2)]$. Hence the amount of time spent in state 1 up through time $(\mu_1 + \mu_2)n$ is approximately

$$n\mu_1 + \sqrt{n}\sigma_1 B_1 - \frac{\mu_1}{\mu_1 + \mu_2}\sqrt{n}(\sigma_1 B_1 + \sigma_2 B_2)$$

$$= n\mu_1 + \frac{\sqrt{n}}{\mu_1 + \mu_2}[\mu_2 \sigma_1 B_1 - \mu_1 \sigma_2 B_2].$$

Since B_1 and B_2 are independent, we can write this as

$$n\mu_1 + \sqrt{n}\sqrt{(\frac{\sigma_1 \mu_2}{\mu_1 + \mu_2})^2 + (\frac{\sigma_2 \mu_1}{\mu_1 + \mu_2})^2}B = \frac{1}{\alpha_1}n + \sqrt{2n}\frac{1}{\alpha_1 + \alpha_2}B,$$

where B is a unit normal. If we let $t = (\mu_1 + \mu_2)n$ we see that the distribution of

$$\frac{Y_t - \frac{\alpha_2}{\alpha_1 + \alpha_2}t}{\bar{\sigma}\sqrt{t}}$$

approaches a unit normal where

$$\bar{\sigma}^2 = \frac{2\alpha_1 \alpha_2}{(\alpha_1 + \alpha_2)^3}.$$

6.2 Renewal Equation

We will be interested in the long-range behavior of renewal processes. Assume we have a renewal process with waiting times $T_1, T_2, \ldots$ with mean μ as defined in the previous section. For $T \geq 0$, we let $U(t)$ be the expected number of occurrences up through time t, where for convenience we will say that an event occurs at time 0. In other words,

$$U(t) = E(N_t + 1).$$

The first renewal theorem states that

$$\lim_{t \to \infty} \frac{U(t)}{t} = \frac{1}{\mu}. \tag{6.3}$$

This is almost a consequence of (6.2); one does need to be a little careful, however, because it is possible for random variables to converge without the expectations converging. We leave the derivation of (6.3) from (6.2) to the exercises (Exercise 6.5).

To analyze the long-term behavior of renewal processes we will need a second, stronger version of the renewal theorem. The second renewal theorem can be thought of as a "derivative" form of (6.3) or as a statement that the renewal process converges to a steady state. The second renewal theorem states that under appropriate hypotheses, for every $r > 0$,

$$\lim_{t \to \infty} U(t+r) - U(t) = \frac{r}{\mu}, \tag{6.4}$$

i.e., for large t, the expected number of renewals in any interval of length r is about r/μ. It is not too difficult to see that some restrictions must be put on the distribution for (6.4) to hold. For example, if the waiting times T_i take on only integer values, then for every integer n,

$$U(n) = U(n + \frac{1}{2}),$$

since renewals occur only at integer times. It turns out that this is really the only thing that can go wrong. We say that a nonnegative random variable X has a lattice distribution if there exists a number a such that with probability 1 the value of X lies in

$$\{ak : k = 0, 1, 2 \ldots\},$$

and we call the smallest such a the period of the distribution. Otherwise we say the X has a nonlattice distribution. We now state the second renewal theorem.

Theorem. If $T_1, T_2, \ldots$ have a nonlattice distribution, then for every $r > 0$,

$$\lim_{t \to \infty} U(t+r) - U(t) = \frac{r}{\mu}.$$

If the $T_1, T_2, \ldots$ have a lattice distribution with period a, then

$$\lim_{n \to \infty} U((n+1)a) - U(na) = \frac{a}{\mu}.$$

We will not give a proof of the nonlattice form of this theorem, but rather will concentrate on showing how it is used. In the next section we will relate the lattice form of this theorem to known results about positive recurrent Markov chains. Let F denote the distribution of T_i. Recall that the convolution of two distributions F, G of nonnegative random variables is defined by

$$F * G(t) = \int_0^t F(t-s) \, dG(s) = \int_0^t G(t-s) \, dF(s).$$

The convolution gives the distribution function of the sum of two independent random variables with distribution functions F and G respectively. Let F be the distribution function for the T_i. We will write $F^{(n)}$ for the convolution of F n times, i.e., for the distribution of $T_1 + \cdots + T_n$. For convenience we will write $F^{(0)}$ for the trivial distribution function associated to the random variable which is identically 0. Recall [see (1.13)] that if Y is a random variable taking values in the nonnegative integers, then

$$E(Y) = \sum_{n=1}^{\infty} P\{Y \geq n\}.$$

Using this, we can write the renewal function $U(t)$ as

$$
\begin{aligned}
U(t) = E(N_t + 1) &= 1 + \sum_{n=1}^{\infty} P\{N_t \geq n\} \\
&= 1 + \sum_{n=1}^{\infty} P\{T_1 + \cdots + T_n \leq t\} \\
&= \sum_{n=0}^{\infty} F^{(n)}(t).
\end{aligned}
$$

Let A_t denote the time elapsed since the last renewal,

$$A_t = \begin{cases} t & \text{if } N_t = 0, \\ t - (T_1 + \cdots + T_n) & \text{if } N_t = n. \end{cases}$$

If we think of the times T_i as being lifetimes of some component, then A_t represents the age of the current component. We would like to determine the steady-state distribution of A_t, i.e., we would like to determine for each x,

$$\Psi_A(x) = \lim_{t \to \infty} P\{A_t \leq x\}.$$

We will condition on the first renewal. One way for A_t to be less than x is for no event to have occurred up through time t and $t \leq x$. This corresponds to $t < T_1$ and has probability $1 - F(t)$ if $t \leq x$. If the first renewal has occurred before time t, at time s say, then the renewal process starts over and there is time $t - s$ left until time t. From this we get the equation

$$P\{A_t \leq x\} = 1_{[0,x]}(t)[1 - F(t)] + \int_0^t P\{A_{t-s} \leq x\} \, dF(s). \qquad (6.5)$$

Here $1_{[0,x]}(t)$ denotes the function that equals 1 for $0 \leq t \leq x$ and equals zero otherwise. If we let $\phi(t) = \phi(t, x) = P\{A_t \leq x\}$, then this becomes

$$\phi(t) = 1_{[0,x]}(t)[1 - F(t)] + \int_0^t \phi(t - s) \, dF(s).$$

This is an example of a renewal equation. We will now consider solutions to renewal equations of the form

$$\phi(t) = h(t) + \int_0^t \phi(t - s) \, dF(s), \qquad (6.6)$$

or in the language of convolutions,

$$\phi(t) = h(t) + \phi * F(t).$$

We will need the associativity property for convolutions: if F and G are distribution functions

$$(\phi * F) * G(t) = \phi * (F * G)(t). \tag{6.7}$$

Let us derive this in the case where F and G have densities, so that $dF(t) = f(t)\,dt$ and $dG(t) = g(t)\,dt$. In this case

$$
\begin{aligned}
(\phi * F) * G(t) &= \int_0^t (\phi * F)(t - s)g(s)\,ds \\
&= \int_0^t [\int_0^{t-s} \phi(t - s - r)f(r)\,dr]\,g(s)\,ds \\
&= \int_0^t [\int_s^t \phi(t - y)f(y - s)\,dy]\,g(s)\,ds \\
&= \int_0^t \phi(t - y)[\int_0^y f(y - s)g(s)\,ds]\,dy \\
&= \int_0^t \phi(t - y)(f * g)(y)\,dy \\
&= \phi * (F * G)(t).
\end{aligned}
$$

We will first show that there is only one solution to (6.6) in the sense that there is at most one $\phi(t)$ that satisfies (6.6) with $\phi(t) = 0$ for $t < 0$ and such that for each t there is a number $M = M_t < \infty$ with $|\phi(s)| \le M$ for all $0 \le s \le t$. Assume there were two such solutions, $\phi_1(t)$ and $\phi_2(t)$, for a given h. Then $\psi(t) = \phi_1(t) - \phi_2(t)$ satisfies $|\psi(s)| \le 2M$, $0 \le s \le t$, and

$$\psi(t) = \int_0^t \psi(t - s)\,dF(s).$$

If we iterate (6.7) we see for each n,

$$\psi(t) = \int_0^t \psi(t - s)\,dF^{(n)}(s).$$

But,

$$|\psi(t)| = |\int_0^t \psi(t - s)\,dF^{(n)}(s)| \le 2M\,F^{(n)}(t).$$

But for fixed t, $F^{(n)}(t) \to 0$ as $n \to \infty$. This shows that $\psi(t) = 0$.

Now that we know there is only one solution, we need only produce a solution. Let

$$\phi(t) = \int_0^t h(t - s)\,dU(s) = \sum_{n=0}^{\infty} \int_0^t h(t - s)\,dF^{(n)}(s)$$

$$= h(t) + \sum_{n=1}^{\infty} \int_0^t h(t-s) \, dF^{(n)}(s).$$

Then one can see, using (6.7), that this satisfies (6.6). This therefore gives the unique solution.

Let us now assume that the F is a nonlattice distribution. Another way of stating the second renewal theorem is to say that for large s,

$$dU(s) \approx \mu^{-1} ds.$$

If $h(t)$ is a bounded function with $\int_0^\infty |h(t)| \, dt < \infty$, then this implies that

$$\lim_{t\to\infty} \int_0^t h(t-s) \, dU(s) = \lim_{t\to\infty} - \int_0^t h(s) \, dU(t-s) = \frac{1}{\mu} \int_0^\infty h(s) \, ds. \quad (6.8)$$

Since the age distribution A_t satisfies (6.5), we can conclude that the long-range age distribution function $\Psi_A(x)$ is given by

$$\Psi_A(x) = \lim_{t\to\infty} P\{A_t \le x\} \quad = \quad \frac{1}{\mu} \int_0^\infty 1_{[0,x]}(s)[1 - F(s)] \, ds$$

$$= \quad \frac{1}{\mu} \int_0^x [1 - F(s)] \, ds.$$

Note that

$$\lim_{x\to\infty} \Psi_A(x) \quad = \quad \frac{1}{\mu} \int_0^\infty [1 - F(s)] \, ds$$

$$= \quad \frac{1}{\mu} \int_0^\infty \int_s^\infty dF(r) \, ds$$

$$= \quad \frac{1}{\mu} \int_0^\infty [\int_0^r ds] \, dF(r)$$

$$= \quad \frac{1}{\mu} \int_0^\infty r \, dF(r) = 1,$$

so this gives a valid distribution function. It has density

$$\psi_A(x) = \Psi_A'(x) = \frac{1}{\mu}[1 - F(x)], \quad 0 < x < \infty.$$

Example 1. Suppose that the waiting times are exponential with rate λ, so that $F(t) = 1 - e^{-\lambda t}$, $\mu = 1/\lambda$. Then

$$\Psi_A(x) = \lim_{t\to\infty} P\{A_t \le x\} = \frac{1}{\mu} \int_0^x \lambda e^{-\lambda s} \, ds = 1 - e^{-\lambda x}.$$

Hence the long-range age distribution for a Poisson process with rate λ is an exponential distribution with rate λ. This is very plausible: at a large time t, the age A_t is the amount of time in the past one must go to see an event. This backwards process looks also like a Poisson process, so the time until an event should be exponential.

Example 2. Suppose that the waiting time distribution is uniform on $[0, 10]$ so that $F(t) = (t/10)$, $0 \le t \le 10$, and $\mu = 5$. Then the age A_t is always less than 10 and for $x < 10$,

$$\Psi_A(x) = \lim_{t \to \infty} P\{A_t \le x\} = \frac{1}{\mu} \int_0^x [1 - \frac{t}{10}] \, dt = \frac{x}{5} - \frac{x^2}{100}. \qquad (6.9)$$

Note in this case (as in essentially all cases but for exponential waiting times) the long-range age distribution is not the same as the waiting time distribution.

We will now consider two other process, the residual life

$$B_t = \inf\{s : N_{t+s} > N_t\},$$

and the total lifetime

$$C_t = A_t + B_t.$$

The residual life gives the amount of time until the current component in a system fails. Consider $P\{B_t \le x\}$. There are two ways for B_t to be less than x. One way is for there to be no renewals up to time t and $B_t \le x$. This corresponds to $t < T_1 \le t + x$ which has probability $F(t + x) - F(t)$. The other possibility is that there is a first renewal at time $s < t$ in which case we need to consider $\{B_{t-s} \le x\}$. This gives the renewal equation

$$P\{B_t \le x\} = [F(t + x) - F(t)] + \int_0^t P\{B_{t-s} \le x\} \, dF(s).$$

The solution to this renewal equation is

$$P\{B_t \le x\} = \int_0^t [F(t - s + x) - F(t - s)] \, dU(s).$$

From (6.8), we can determine the long-range residual life distribution function $\Psi_B(x)$,

$$
\begin{aligned}
\Psi_B(x) &= \lim_{t \to \infty} \int_0^t [F(t - s + x) - F(t - s)] \, dU(s) \\
&= \lim_{t \to \infty} - \int_0^t [F(s + x) - F(s)] \, dU(t - s) \\
&= \frac{1}{\mu} \int_0^\infty [F(s + x) - F(s)] \, ds \\
&= \frac{1}{\mu} [\int_0^\infty [1 - F(s)] \, ds - \int_0^\infty [1 - F(s + x)] \, ds] \\
&= \frac{1}{\mu} [\int_0^\infty [1 - F(s)] \, ds - \int_x^\infty [1 - F(r)] \, dr] \\
&= \frac{1}{\mu} \int_0^x [1 - F(s)] \, ds.
\end{aligned}
$$

What we see is that the long time distribution function for the residual life is the same as that for the age distribution. If one thinks about this, it is reasonable. Consider every lifetime T_i. For every r, s with $r + s = T_i$, there will correspond one time t when $A_t = r$, $B_t = s$ and another time u when $A_u = s$, $B_u = r$. By this symmetry, we would expect A_t and B_t to have the same limiting distribution.

Now consider the total lifetime C_t and $P\{C_t \le x\}$. One way for C_t to be less than x is for there to be no renewals up through time t and the total lifetime less than x. This corresponds to $t < T_1 \le x$ which has probability $F(x) - F(t)$. The other possibility is that the first event occurs at some $s < t$ in which case we need to consider $P\{C_{t-s} \le x\}$. This gives the renewal equation

$$P\{C_t \le x\} = 1_{[0,x]}(t)[F(x) - F(t)] + \int_0^t P\{C_{t-s} \le x\} \, dF(s).$$

By solving the renewal equation and using (6.8), we see that the limiting distribution for the lifetime, $\Psi_C(x)$ is given by

$$
\begin{aligned}
\Psi_C(x) &= \lim_{t \to \infty} \int_0^t 1_{[0,x]}(t - s)[F(x) - F(t - s)] \, dU(s) \\
&= \lim_{t \to \infty} - \int_0^t 1_{[0,x]}(s)[F(x) - F(s)] \, dU(t - s) \\
&= \frac{1}{\mu} \int_0^\infty 1_{[0,x]}(s)[F(x) - F(s)] \, ds \\
&= \frac{1}{\mu} \int_0^x [F(x) - F(s)] \, ds \\
&= \frac{1}{\mu} [xF(x) - \int_0^x F(s) \, ds].
\end{aligned}
$$

This formula is best understood in the case where F has a density $f(t)$. In this case $\Psi_C(x)$ has density

$$\psi_C(x) = \Psi'_C(x) = \frac{1}{\mu} x f(x). \tag{6.10}$$

This can be understood intuitively. Suppose $x < y$. Then the relative "probability" of waiting times of size x and size y is $f(x)/f(y)$. However, every waiting time of size y uses up y units of time while a waiting time of size x uses up x units of time. So the ratio of times in an interval of size x to an interval of size y should be $xf(x)/yf(y)$. The $1/\mu$ can easily be seen to be the appropriate normalization factor to make this a probability density.

Example 3. If the waiting times are exponential with rate λ, then $\mu = 1/\lambda$ and Ψ_A and Ψ_B have exponential distributions with rate λ. Note that

Ψ_C has density
$$\psi_C(x) = \lambda^2 x e^{-\lambda x}.$$

This is the density of a Gamma distribution with parameters 2 and λ and is the density of the sum of two independent exponential random variables with rate λ. For large time, the age and the residual life are independent random variables.

Example 4. If F is uniform on $[0, 10]$, then $\mu = 5$, and Ψ_A and Ψ_B are given by (6.9) with densities

$$\psi_A(x) = \psi_B(x) = \frac{1}{5} - \frac{x}{50}, \quad 0 < x < 10.$$

Note that the expected age or the expected residual life in the long run is given by

$$\int_0^{10} x[\frac{1}{5} - \frac{x}{50}] \, dx = \frac{10}{3}.$$

The density of Ψ_C is given by

$$\psi_C(x) = \frac{1}{\mu} x f(x) = \frac{x}{50}, \quad 0 < x < 10.$$

It is easy to check that the age and residual life are not asymptotically independent in this case, e.g., there is a positive probability that the age is over 8 and a positive probability that the residual life is over 8, but it is impossible for both of them to be over 8 since the total lifetime is bounded by 10.

Suppose one is replacing components as they fail and the lifetimes are independent with distribution F. Suppose we consider the system at some large t, and ask how long the present component is expected to last. This is equivalent to finding the expected value of the residual life. This is given by

$$\int_0^\infty x \psi_B(x) \, dx = \frac{1}{\mu} \int_0^\infty x[1 - F(x)] \, dx = \frac{1}{2\mu} \int_0^\infty x^2 \, dF(x).$$

The last equality is obtained by integrating by parts. It is easy to give examples (see Exercise 6.6) of distributions of densities $f(x)$ such that

$$\mu < \frac{1}{2\mu} \int_0^\infty x^2 f(x) \, dx.$$

In fact, it is possible for $\mu < \infty$ and the expected residual lifetime to be infinite. This may be surprising at first; however, a little thought will show that this is not so unreasonable.

We finish this section by describing how to create a "stationary renewal process." Suppose $T_1, T_2, \ldots$ are independent with nonlattice distribution F. Let Ψ_B be the long-range residual life distribution and let Y be a random variable independent of $T_1, T_2, \ldots$ with distribution function Ψ_B. Define N_t as in (6.1). Then N_t looks like a renewal process in steady state. It has the property that for every $s < t$, $N_t - N_s$ has the same distribution as N_{t-s}.

6.3 Discrete Renewal Processes

In this section we will suppose that the random variables $T_1, T_2, \ldots$ are lattice random variables. Without loss of generality we will assume that the period a as defined in Section 6.2 is equal to 1 (the period is always equal to 1 in some choice of time units). Let F be the distribution function for the T_i and let

$$p_n = P\{T_i = n\} = F(n) - F(n-1).$$

We will assume for ease that $p_0 = 0$; if $p_0 > 0$ we can make a slight adjustment of the methods in this section (see Exercise 6.7). Since the period is 1, the greatest common divisor of the set

$$\{n : p_n > 0\}$$

is 1. As before set

$$\mu = E(T_i) = \sum_{n=1}^{\infty} n p_n,$$

and we assume $\mu < \infty$.

Let N_j denote the number of events that have occurred up through (and including) time j, i.e., $N_j = 0$ if $j < T_1$ and otherwise

$$N_j = \max\{n : T_1 + \cdots + T_n \leq j\}.$$

We can also define the age process A_j by $A_j = j$ if $j < T_1$ and otherwise

$$A_j = j - (T_1 + \cdots + T_{N_j}).$$

The key fact is that A_j is a Markov chain. Let

$$\lambda_n = P\{T_i = n \mid T_i > n-1\} = \frac{p_n}{1 - F(n-1)}.$$

Then A_j is a discrete-time Markov chain with transition probabilities

$$p(n,0) = \lambda_{n+1}, \quad p(n, n+1) = 1 - \lambda_{n+1}.$$

Let K be the largest number k such that $p_k > 0$ (where $K = \infty$ if $p_k > 0$ for infinitely many k). Then A_j is an irreducible Markov chain with state space $\{0, 1, \ldots, K-1\}$ if $K < \infty$ and state space $\{0, 1, \ldots\}$ if $K = \infty$. The chain is also aperiodic since we assumed the period of F is 1. We start with $A_0 = 0$ and note that the nth return to state 0 occurs at time $T_1 + \cdots + T_n$. The condition $E(T_i) < \infty$ implies that A_j is a positive recurrent chain.

The invariant measure π for this chain can be obtained by solving the equations

$$\pi(n+1) = p(n, n+1)\pi(n) = (1 - \lambda_{n+1})\pi(n)$$
$$= \frac{1 - F(n+1)}{1 - F(n)}\pi(n), \quad n > 0,$$

$$\pi(0) = \sum_{n=0}^{\infty} p(n, 0)\pi(n) = \sum_{n=0}^{\infty} \lambda_{n+1}\pi(n).$$

The first equations can be solved recursively to yield

$$\pi(n) = [1 - F(n)]\pi(0).$$

To find the value for $\pi(0)$ for which $\sum \pi(n) = 1$, we check that

$$\sum_{n=0}^{\infty}[1 - F(n)] = \sum_{n=0}^{\infty} \sum_{m=n+1}^{\infty} p_m$$
$$= \sum_{m=1}^{\infty} p_m \sum_{n=0}^{m-1} 1$$
$$= \sum_{m=1}^{\infty} m p_m = \mu.$$

In particular,

$$\pi(0) = \frac{1}{\mu}.$$

Note that

$$P\{\text{an event at time } j\} = P\{N_j > N_{j-1}\} = P\{A_j = 0\}.$$

Since A_j is an aperiodic, irreducible, positive recurrent Markov chain we know that

$$\lim_{j \to \infty} P\{A_j = 0\} = \pi(0) = \frac{1}{\mu}.$$

This gives the improved renewal theorem for discrete renewal processes.

We have also derived the long time age distribution,

$$\psi_A(n) = \lim_{j \to \infty} P\{A_j = n\} = \pi(n) = \frac{1 - F(n)}{\mu}.$$

Consider the residual life,

$$B_j = \min\{k > 0 : N_{j+k} > N_j\}.$$

We can compute the long time distribution of B_j,

$$\psi_B(n) = \lim_{j \to \infty} P\{B_j = n\}$$

$$= \lim_{j \to \infty} \sum_{m=0}^{\infty} P\{A_j = m\} P\{B_j = n \mid A_j = m\}$$

$$= \sum_{m=0}^{\infty} \pi(m) P\{B_j = n \mid A_j = m\}$$

$$= \sum_{m=0}^{\infty} \frac{1 - F(m)}{\mu} \frac{p_{n+m}}{1 - F(m)}$$

$$= \frac{1}{\mu} \sum_{m=0}^{\infty} p_{n+m}$$

$$= \frac{1 - F(n-1)}{\mu}.$$

In other words,

$$\psi_B(n) = \lim_{j \to \infty} P\{B_j = n\} = \lim_{j \to \infty} P\{A_j = n - 1\} = \psi_A(n-1).$$

The residual life has the same long time distribution as the age except for a difference of 1 which comes from the fact that the smallest value for the residual life is 1 while the smallest value for the age is 0. For the total lifetime of the component at time j,

$$C_j = A_j + B_j,$$

we can compute

$$\psi_C(n) = \lim_{j \to \infty} P\{C_j = n\}$$

$$= \lim_{j \to \infty} \sum_{m=0}^{n-1} P\{A_j = m\} P\{C_j = n \mid A_j = m\}$$

$$= \sum_{m=0}^{n-1} \pi(m) P\{C_j = n \mid A_j = m\}$$

$$= \sum_{m=0}^{n-1} \frac{1 - F(m)}{\mu} \frac{p_n}{1 - F(m)}$$

$$= \frac{1}{\mu} \sum_{m=0}^{n-1} p_n$$

$$= \frac{n p_n}{\mu}.$$

This is the discrete analogue of (6.10).

Example 1. Bernoulli Process. The discrete analogue of the Poisson process is the Bernoulli process. Let $0 < p < 1$ and let $X_1, X_2, \ldots$ be independent random variables with $P\{X_i = 1\} = 1 - P\{X_i = 0\} = p$.

$N_j = X_1 + \cdots + X_j$ represents the number of "successes" in the first j trials of an experiment with probability p of success. The waiting times T_i have a geometric distribution

$$P\{T_i = n\} = (1 - p)^{n-1}p, \quad n \geq 1,$$

with $\mu = 1/p$. The asymptotic age distribution is given by

$$\psi_A(n) = \frac{1 - F(n)}{\mu} = p \sum_{j=m+1}^{\infty} (1 - p)^{m-1}p = p(1 - p)^n,$$

i.e., the age is one less than a random variable with a geometric distribution. The residual life distribution is geometric with parameter p. The asymptotic lifetime distribution is given by

$$\phi_C(n) = np^2(1 - p)^{n-1},$$

which is the distribution of the sum of two independent random variables with distributions ϕ_A and ϕ_B, respectively. The age and the residual life are independent.

Example 2. Suppose F is uniformly distributed on $\{1, \ldots, 10\}$ with $\mu = 11/2$. Then

$$F(n) = \frac{n}{10}, \quad n = 1, 2, \ldots, 10.$$

The asymptotic age distribution is given by

$$\psi_A(n) = \frac{1 - F(n)}{\mu} = \frac{10 - n}{55}, \quad n = 0, \ldots, 9.$$

and for large time the residual life distribution is given by

$$\psi_B(n) = \frac{1 - F(n - 1)}{\mu} = \frac{11 - n}{55}, \quad n = 1, \ldots, 10.$$

The asymptotic lifetime distribution is given by

$$\psi_C(n) = \frac{n}{55}, \quad n = 1, 2, \ldots, 10.$$

In this case, the age and residual life are not asymptotically independent.

6.4 $M/G/1$ and $G/M/1$ **Queues**

We will consider Example 4 from Section 6.1. Customers arrive into a single-server queue from a Poisson Process with rate λ. Customers are served (first come, first served) and the service time is a random variable with distribution function F and mean $\mu < \infty$. We will call the service rate $1/\mu$, even though the service times are not exponential. The service times and the arrival times are independent. As mentioned before there is a natural renewal process involved where $R_1, R_2, \ldots$ denote the amount of time spent in "idle times" while $S_1, S_2, \ldots$ denote the amount of time spent in "busy times." If the queue starts idle, i.e., if $X_0 = 0$ where X_t denotes the size of the queue (including the person being served) at time t, then the time until the start of the next idle time is given by $T_1 = R_1 + S_1$ and the time until the start of the $(n + 1)$st idle time is given

$$T_1 + \cdots + T_n,$$

where $T_i = R_i + S_i$.

The times R_i are exponential with rate λ, i.e, with mean $1/\lambda$. The distribution of the S_i is more difficult to determine. However, we will be able to determine $E(S_i)$. Assume that the service rate is greater than the arrival rate, i.e.,

$$\mu\lambda < 1.$$

Consider the start of a busy time, so that $X_t = 1$. We will consider a discrete-time Markov chain Y_n that gives the number of people in the queue immediately after the nth person has been served. We start with $Y_0 = 1$. The value Y_1 is obtained by considering the number of people who entered the queue during the first service time and subtracting 1 (for the person who has left the queue). For $i > 1$, Y_i is obtained by adding to Y_{i-1} the number of people who entered the queue while the ith person was being serviced and subtracting one. Let

$$\tau = \min\{n : Y_n = 0\}.$$

If $U_1, U_2, \ldots$ denote the service times of the customers, then the length of the first busy time is given by

$$S_1 = U_1 + U_2 + \cdots + U_\tau.$$

The $U_1, U_2, \ldots$ are indepedent random variables, each with distribution function F, but the U_i are not independent of τ. Still we might hope to conclude

$$E(S_1) = E(U_i)E(\tau). \tag{6.11}$$

In fact, this equation is valid assuming $E(\tau) < \infty$. This follows from Wald's equation which was proved in Exercise 5.14. It was shown in Exercise 5.12 that $E(\tau) < \infty$ if $E(Y_i) < 0$ and in this case another application of Wald's equation can be made to show that

$$E(\tau) = -\frac{1}{E(Y_i)}.$$

Let us compute $E(Y_i)$. The probability that k people arrive in the queue during a service time U_i is

$$q_k = \int_0^\infty P\{k \text{ arrive} \mid U_i = s\}\, dF(s)$$

$$= \int_0^\infty \frac{e^{-s\lambda}(s\lambda)^k}{k!} \, dF(s).$$

The expected number of arrivals is therefore

$$\sum_{k=0}^\infty k q_k = \int_0^\infty \sum_{k=0}^\infty k \frac{e^{-s\lambda}(s\lambda)^k}{k!} \, dF(s)$$

$$= \int_0^\infty s\lambda \, dF(s) = \lambda\mu.$$

Hence $E(Y_i) = \lambda\mu - 1$ and

$$E(\tau) = \frac{1}{1 - \lambda\mu} = \frac{\rho}{\rho - \lambda},$$

where we write $\rho = 1/\mu$ for the service rate. The expected length of a busy time is given by

$$E(S_1) = E(U_i)E(\tau) = \frac{1}{\rho - \lambda}.$$

The fraction of time that the queue is busy is given by

$$\frac{E(S_1)}{E(R_i) + E(S_1)} = \frac{\lambda}{\rho}.$$

Note that this ratio tends to 1 as $\lambda \to \rho$.

If $\lambda = \rho$, the chain Y_n can be shown to be recurrent (see Exercise 2.10) so that the queue size returns to 0 infinitely often. However, in the long run the fraction of time spent with the queue empty goes to 0. If $\lambda > \rho$, the chain Y_n is transient, and hence the queue size goes to infinity.

Now let us consider the somewhat less realistic $G/M/1$ queue. Here customers arrive one at a time with waiting times $T_1, T_2, \ldots$ having common distribution function F with mean $1/\lambda$. There is one server and the service times are exponential with rate ρ. We will assume that the service rate is greater than the arrival rate, $\rho > \lambda$.

There exists a natural Markov chain embedded in the $G/M/1$ queue. Consider Y_n the number of customers in the system immediately before the nth customer arrives. (We will assume that the queue starts out empty and we set $Y_0 = 0$.) Then Y_n can easily be checked to be a Markov chain with state space $\{0, 1, 2, \ldots\}$.

To compute the transition probability for this chain we first for ease consider what happens if there are an infinite number of people in the queue. Let q_k be the probability that exactly k individuals are served between the arrival times of two successive customers. If the arrival time is t, then the number of customers served has a Poisson distribution with parameter ρt. Hence

$$q_k = \int_0^\infty P\{k \text{ served} \mid T_i = t\} \, dF(t)$$

$$= \int_0^\infty e^{-\rho t} \frac{(\rho t)^k}{k!} \, dF(t).$$

The expected number served is

$$\sum_{k=0}^\infty k q_k = \sum_{k=0}^\infty k \int_0^\infty e^{-\rho t} \frac{(\rho t)^k}{k!} \, dF(t)$$

$$= \int_0^\infty [\sum_{k=0}^\infty k e^{-\rho t} \frac{(\rho t)^k}{k!}] \, dF(t)$$

$$= \int_0^\infty \rho t \, dF(t)$$

$$= \rho/\lambda > 1.$$

Now if $Y_n = j$, then after the nth customer arrives there will be $j+1$ customers in the queue. The queue will serve customers until the queue empties. It is easy to see then that

$$P\{Y_{n+1} = k \mid Y_n = j\} = q_{(j+1)-k}, \quad k > 0,$$

$$P\{Y_{n+1} = 0 \mid Y_n = j\} = \sum_{k \leq 0} q_{(j+1)-k}.$$

If we set $p_k = q_{1-k}$, the we see that Y_n has transition probabilities

$$p(j,k) = p_{k-j}, \quad k > 0,$$

$$p(j,0) = \sum_{k \leq 0} p_{k-j}.$$

It can be shown (see Exercise 2.11) that this is a positive recurrent Markov chain. Its invariant probability is of the form

$$\pi(j) = \beta^j (1 - \beta),$$

where β is the unique solution to

$$\beta = \sum_{j=0}^\infty q_j \beta^j,$$

with $\beta \in (0,1)$. It is hard to evaluate β analytically but it can be computed numerically.

6.5 Exercises

6.1 Suppose the lifetime of a component T_i in hours is uniformly distributed on $[100, 200]$. Components are replaced as soon as one fails and assume that this process has been going on long enough to reach equilibrium.

(a) What is the probability that the current component has been in operation for at least 50 hours?

(b) What is the probability that the current component will last for at least 50 more hours?

(c) What is the probability that the total lifetime of the current component will be at least 150 hours?

(d) Suppose it is known that the current component has been in operation for exacty 90 hours. What is the probability that it will last at least 50 more hours?

6.2 Repeat Exercise 6.1 with the T_i exponentially distributed with mean 150.

6.3 Repeat Exercise 6.1 with the T_i having density

$$f(t) = \frac{1}{t \ln 2}, \quad 100 < t < 200.$$

6.4 Repeat Exercise 6.1 with the T_i having distribution

$$P\{T_i = 100\} = P\{T_i = 200\} = 1/2.$$

6.5 Let N_t denote the renewal process associated with independent, identically distributed random variables $T_1, T_2, \ldots$ with mean μ.

(a) Explain why for any positive integers j, k and any t, the following inequality holds

$$P\{N_t \geq jk\} \leq [P\{N_t \geq j\}]^k.$$

(b) The law of large numbers for renewal processes, (6.2), states that for every $\epsilon > 0$

$$\lim_{t \to \infty} P\{\frac{t(1-\epsilon)}{\mu} \leq N_t \leq \frac{t(1+\epsilon)}{\mu}\} = 1. \tag{6.12}$$

Use (a) and (6.12) to conclude that for every $\epsilon > 0$,

$$\lim_{t \to \infty} \frac{1}{t} E(N_t I\{N_t > \frac{t(1+\epsilon)}{\mu}\}) = 0.$$

(c) Derive the first renewal theorem, (6.3).

6.6 Assume that the waiting times T_i have distribution

$$P\{T_i = 1\} = \frac{9}{10}, \quad P\{T_i = 10\pi\} = \frac{1}{10}.$$

Note that the times T_i have a nonlattice distribution.

(a) What is the age distribution $\Psi_C(n)$?

(b) For large times, what is the expected residual life? Compare to $E(T_i)$.

6.7 Assume $T_1, T_2, \ldots$ are independent identically distributed nonnegative random variables with $P\{T_i = 0\} = q \in (0, 1)$. Suppose the distribution function of the T_i is F with mean μ, and let G be the conditional distribution function of the T_i given that the $T_i > 0$,

$$G(x) = P\{T_i \le x \mid T_i > 0\} = \frac{F(x) - F(0)}{1 - q}.$$

Let $\tilde{T}_1, \tilde{T}_2, \ldots$ be independent, identically distributed random variables with distribution function G and let $U(t)$ and $\tilde{U}(t)$ be the renewal functions associated with the T_i and the $\tilde{T}_i$ respectively. Show that

$$\tilde{U}(t) = (1 - q)U(t).$$

CHAPTER 7

Reversible Markov Chains

7.1 Reversible Processes

In this chapter we will study a particular class of Markov chains, reversible chains. A large number of important chains are reversible, and we can take advantage of this fact in trying to understand their behavior.

Suppose we have a continuous-time Markov chain X_t taking values in state space S (finite or countably infinite) with transition rates $\alpha(x, y)$. If π is any measure on S, i.e., a nonnegative function on S, then the chain is said to be *reversible with respect to the measure* π if for all $x, y \in S$,

$$\pi(x)\alpha(x, y) = \pi(y)\alpha(y, x).$$

We will say that the chain is *symmetric* if for every x, y

$$\alpha(x, y) = \alpha(y, x).$$

Note that a chain is symmetric if and only if it is reversible with respect to the uniform measure $\pi(x) = 1$, $x \in S$. Similarly, a discrete-time Markov chain with transition matrix P is said to be reversible with respect to π if

$$\pi(x)P(x, y) = \pi(y)P(y, x),$$

for all $x, y \in S$ and symmetric if $P(x, y) = P(y, x)$. In the next two sections we will discuss continuous-time chains, but analogous statements hold for discrete-time chains.

Example 1. Let $G = (V, E)$ be a graph as in Chapter 1, Example 5. Let $S = V$ and

$$\alpha(x, y) = 1/d(x), \ (x, y) \in E,$$

where $d(x)$ is the number of vertices adjacent to x. This is a continuous-time analogue of Example 5. Then this chain is reversible with respect to the measure $\pi(x) = d(x)$. If instead we choose

$$\alpha(x, y) = 1, \ (x, y) \in E,$$

then the chain is symmetric and hence reversible with respect to the uniform measure.

Example 2. Let $G = (V, E)$ be any graph and let $g : E \to [0, \infty)$. Such a configuration is often called a network. A network gives rise to a symmetric chain with transitions

$$\alpha(x, y) = \alpha(y, x) = g(e),$$

if e denotes the edge connecting x and y.

Example 3. Suppose we have a birth-and-death chain on $S = \{0, 1, 2, \ldots\}$ with birth rates λ_n and death rates μ_n. In other words, the transition rates are

$$\alpha(n, n+1) = \lambda_n, \quad \alpha(n, n-1) = \mu_n.$$

Let $\pi(0) = 1$ and for $n > 0$

$$\pi(n) = \frac{\lambda_0 \lambda_1 \cdots \lambda_{n-1}}{\mu_1 \mu_2 \cdots \mu_n}.$$

Then the chain is reversible with respect to the measure π.

Example 4. Let $G = (V, E)$ be any graph and suppose $\pi : V \to (0, \infty)$ is a positive measure on G. Suppose each vertex is adjacent to only a finite number of other vertices. Define $\alpha(x, y) = 0$ if (x, y) is not an edge of G and for $(x, y) \in E$,

$$\alpha(x, y) = \min\{1, \frac{\pi(y)}{\pi(x)}\}.$$

Then α generates a chain that is reversible with respect to π.

If a chain is reversible with respect to π, then

$$\sum_{y \in S} \pi(y)\alpha(y, x) = \pi(x) \sum_{y \in S} \alpha(x, y) = \pi(x)\alpha(x),$$

i.e., π is an invariant measure for α. If the state space is finite, or if the state space is infinite with $\sum \pi(x) < \infty$, then we can normalize π so that it is an invariant probability for α. In particular, if α is irreducible, we know that if α is reversible with respect to a probability measure π then π is the (unique) invariant measure. Conversely, if an irreducible chain is reversible with respect to a π with $\sum \pi(x) = \infty$, we can conclude that there is no invariant probability measure and hence the chain is null recurrent or transient.

What the reversibility condition states is that the system, in equilibrium, looks the same whether time goes forward or backward. To give an easy example of a nonreversible chain consider the three-state chain on $S = \{0, 1, 2\}$ with rates

$$\alpha(0, 1) = \alpha(1, 2) = \alpha(2, 0) = 1,$$

$$\alpha(1, 0) = \alpha(2, 1) = \alpha(0, 2) = 2.$$

This is clearly irreducible with invariant probability measure $\pi(0) = \pi(1) = \pi(2) = 1/3$. If the chain were to be reversible, it would need to be reversible with respect to π, but clearly

$$\pi(0)\alpha(0,1) \neq \pi(1)\alpha(1,0).$$

7.2 Convergence to Equilibrium

For reversible chains, we can give estimates for the amount of time needed for the chain to reach a measure close to the invariant probability measure. Let X_t be an irreducible continuous-time Markov chain with rates $\alpha(x,y)$, reversible with respect to the probability measure π. We will assume that the state space is finite, $S = \{1, \ldots, N\}$, but one can generalize these ideas to positive recurrent chains on an infinite state space. For ease, we will only consider the case where A is symmetric (reversible with respect to the uniform measure), but these ideas hold for all reversible chains.

If A is a symmetric matrix, then it can be shown (see an advanced book on linear algebra) that there is a complete set of eigenvalues and eigenvectors. Moreover, all the eigenvalues are real so we can write the eigenvalues in decreasing order,

$$0 = \lambda_1 > \lambda_2 \geq \lambda_3 \geq \cdots \geq \lambda_N.$$

We know $\lambda_2 < 0$ because the chain is irreducible. By symmetry, we see that if $\langle \cdot, \cdot \rangle$ denotes inner product,

$$\langle A\bar{v}, \bar{w} \rangle = \langle \bar{v}, A\bar{w} \rangle = \sum_{i=1}^{N} \sum_{j=1}^{N} v^i w^j A(i,j). \tag{7.1}$$

A matrix satisfying the first equality is said to be self-adjoint (with respect to the uniform measure) and the expression on the right is often called the quadratic form associated with the matrix.

Let

$$\bar{1} = \bar{v}_1, \bar{v}_2, \ldots, \bar{v}_N,$$

be the eigenvectors for A, which are both right and left eigenvectors since A is symmetric. Using (7.1) we can see that

$$\lambda_j \langle \bar{v}_j, \bar{v}_k \rangle = \langle A\bar{v}_j, \bar{v}_k \rangle = \langle \bar{v}_j, A\bar{v}_k \rangle = \lambda_k \langle \bar{v}_j, \bar{v}_k \rangle,$$

and hence eigenvectors for different eigenvalues are orthogonal ($\langle \bar{v}_j, \bar{v}_k \rangle = 0$). We can therefore choose the eigenvectors so they are all orthogonal. These eigenvectors are also the eigenvectors for the matrix e^{tA} with corresponding eigenvalues $e^{t\lambda_j}$,

$$e^{tA}\bar{v}_j = e^{t\lambda_j}\bar{v}_j.$$

Let $U \subset R^N$ denote the $N-1$ dimensional subspace generated by the vectors $\{\bar{v}_2, \ldots, \bar{v}_n\}$, or equivalently, the set of vectors $\bar{w}$ satisfying

$$\sum_{i=1}^{N} w^i = 0.$$

By writing any $\bar{w} \in U$ as a linear combination of left eigenvectors, we can easily see that

$$\|\bar{w}e^{tA}\| \leq e^{t\lambda_2}\|\bar{w}\|,$$

where $\|w\|^2 = \sum_{i=1}^{N}[w^i]^2$. Now suppose we start the chain with any probability vector $\bar{\nu}$. We can write

$$\bar{\nu} = \bar{\pi} + \bar{w},$$

where $\bar{\pi} = (1/N)\bar{1}$ is the invariant probability and $\bar{w} = \bar{\nu} - \bar{\pi} \in U$. Since $\bar{\pi}e^{tA} = \bar{\pi}$, we can conclude

$$\|\bar{\nu}e^{tA} - \bar{\pi}\| = \|(\bar{\nu} - \bar{\pi})e^{tA}\| \leq e^{t\lambda_2}\|\bar{\nu} - \bar{\pi}\|.$$

What we see is that the rate of convergence is essentially controlled by the size of λ_2, and if we can get lower bounds on $|\lambda_2|$, we can bound the rate of convergence.

Example 1. Consider simple random walk on the complete graph on N vertices, i.e., the chain with state space $S = \{1, \ldots, N\}$ and rates

$$\alpha(i, j) = \frac{1}{N-1}, \quad i \neq j.$$

For any i the vector $\bar{v}$ with

$$v^j = \begin{cases} N-1, & i = j \\ -1, & i \neq j, \end{cases}$$

is a right eigenvector with eigenvalue $-N/(N-1)$. These vectors form an $N-1$ dimensional subspace. Hence $\lambda_1 = 0$ and

$$\lambda_2 = \lambda_3 = \cdots = \lambda_N = -\frac{N}{N-1}.$$

For large N, λ_2 is about -1, and the time to equilibrium does not grow as N gets large. This is fairly clear in this example since the process very quickly "forgets" what its initial state was.

Example 2. Consider simple random walk on a circle, i.e., the chain with state space $S = \{1, \ldots, N\}$ and rates $\alpha(x, y) = 1/2$ if $|x - y| = 1 (\mathrm{mod}\ N)$. This is reversible with respect to the uniform measure on S. The eigenvalues for A can be found exactly in this case (see Exercise 7.3),

$$\lambda_j = \cos(\frac{(j-1)2\pi}{N}) - 1, \quad j = 1, 2, \ldots, N.$$

In particular, $\lambda_2 = \cos(2\pi/N) - 1$ which for large N (by the Taylor series for cosine) looks like $-2\pi^2 N^{-2}$. This says that it takes on the order of about N^2 time units in order for the distribution to be within e^{-1} of the uniform distribution. It makes sense that it takes on order N^2 steps to get close to equilibrium, if we remember that it takes a random walker on the order of N^2 steps to go a distance of about N.

3. Let the state space S be all binary sequences of length N, i.e., all N-tuples $(a_1, \ldots, a_N)$, $a_i \in \{0, 1\}$. Note that the state space has 2^N elements. Consider the chain with $\alpha(x, y) = 1$ if x and y are two sequences that differ in exactly one component and $\alpha(x, y) = 0$ otherwise. This is sometimes called random walk on the N-dimensional hypercube. Clearly this is reversible with respect to the uniform measure. It can be shown that $-2j/N$ is an eigenvalue with multiplicity $\binom{N}{j}$. In this case, $\lambda_2 = -2/N$. and it takes on order N steps to get close to equilibrium. This can be understood intuitively by noting that if the number of steps is of order N, most components have had an opportunity to change at least once.

Now let U be the set of vectors that are orthogonal to $\bar{1}$, i.e., the set of vectors $\bar{w}$ satisfying

$$\sum_{i=1}^{N} w^i = 0.$$

If $\bar{w} \in U$, then $A\bar{w} \in U$. If we write

$$\bar{w} = a_2 \bar{v}_2 + \cdots + a_n \bar{v}_n,$$

with $a_i = \langle \bar{v}_i, \bar{w} \rangle$, we see that

$$
\begin{aligned}
\langle \bar{w}, A\bar{w} \rangle &= \sum_{i=2}^{N} \sum_{j=2}^{N} \langle a_i \bar{v}_i, a_j A \bar{v}_j \rangle \\
&= \sum_{i=2}^{N} \sum_{j=2}^{N} a_i a_j \lambda_j \langle \bar{v}_i, \bar{v}_j \rangle \\
&= \sum_{i=2}^{N} a_i^2 \lambda_i \langle \bar{v}_i, \bar{v}_i \rangle \\
&\leq \lambda_2 \sum_{i=2}^{n} \langle a_i \bar{v}_i, a_i \bar{v}_i \rangle = \lambda_2 \langle \bar{w}, \bar{w} \rangle.
\end{aligned}
$$

Also, we get equality in the above expression if we choose $\bar{w} = \bar{v}_2$. What we have derived is the Rayleigh–Ritz variational formulation for the second eigenvalue,

$$\lambda_2 = \sup \frac{\langle \bar{w}, A\bar{w} \rangle}{\langle \bar{w}, \bar{w} \rangle},$$

where the supremum is taken over all vectors $\bar{w}$ with

$$\langle \bar{1}, \bar{w} \rangle = \sum_{i=1}^{N} w^i = 0.$$

Lower bounds for λ_2 (i.e., upper bounds of $|\lambda_2|$) can be obtained by considering particular $\bar{w} \in U$. If $T \subset S$, let $\bar{w} \in U$ with components

$$w^i = \begin{cases} 1 - \pi(T), & i \in T \\ -\pi(T), & i \notin T. \end{cases},$$

where

$$\pi(T) = \frac{\text{number of elements in } T}{N}.$$

Note that $\langle \bar{1}, \bar{w} \rangle = 0$ and

$$\begin{aligned}
\langle \bar{w}, \bar{w} \rangle &= \sum_{i \in T} [1 - \pi(T)]^2 + \sum_{i \notin T} \pi(T)^2 \\
&= [1 - \pi(T)]^2 N \pi(T) + \pi(T)^2 N [1 - \pi(T)] = N \pi(T)[1 - \pi(T)].
\end{aligned}$$

If $i \in T$,

$$\begin{aligned}
(A\bar{w})^i &= \sum_{j} A_{ij} w^j \\
&= -\alpha(i)[1 - \pi(T)] + \sum_{j \in T, j \neq i} \alpha(j, i)[1 - \pi(T)] - \sum_{j \notin T} \alpha(j, i)\pi(T) \\
&= -\sum_{j \notin T} \alpha(j, i)[1 - \pi(T)] - \sum_{j \notin T} \alpha(j, i)\pi(T) \\
&= -\sum_{j \notin T} \alpha(j, i).
\end{aligned}$$

Similarly, if $i \notin T$,

$$(Aw)^i = \sum_{j \in T} \alpha(j, i).$$

Therefore,

$$\begin{aligned}
\langle \bar{w}, A\bar{w} \rangle &= \sum_{i} w^i (A\bar{w})^i \\
&= \sum_{i \in T}[1 - \pi(T)] \sum_{j \notin T}[-\alpha(j, i)] + \sum_{i \notin T}[-\pi(T)] \sum_{j \in T} \alpha(j, i) \\
&= -\sum_{i \in T} \sum_{j \notin T} \alpha(j, i).
\end{aligned}$$

Define κ by

$$\kappa = \inf_{T \subset S} \frac{\sum_{i \in T} \sum_{j \notin T} \alpha(i, j)\pi(i)}{\pi(T)[1 - \pi(T)]}.$$

Then by considering this choice of $\bar{w}$ in the Rayleigh–Ritz formulation, we have

$$|\lambda_2| \leq \kappa.$$

Unfortunately this bound is often not very good. A large area of research is concerned with finding better ways to estimate λ_2; we do not discuss this any further in this book.

7.3 Markov Chain Algorithms

A recent application of Markov chain theory has been in Monte Carlo simulations of random systems. The idea of Monte Carlo simulations is simple: to understand a random system one does many trials on a computer and sees how it behaves. These simulations always use a random number generator, generally a function that produces independent numbers distributed uniformly between 0 and 1. (Actually, a computer can only produce pseudo-random numbers and there are important questions as to whether pseudo-random number generators are "random" enough. We will not worry about that question here and will just assume that we have a means to generate independent identically distributed numbers $U_1, U_2, \ldots$ distributed uniformly on $[0, 1]$.)

As an example, suppose we were interested in studying properties of "random" matrices whose entries are 0s and 1s. As a probability space we could choose the set S of $N \times N$ matrices M, with

$$M(i, j) = 0 \text{ or } 1, \quad 1 \leq i, j \leq N.$$

A natural probability measure would be the uniform measure on all 2^{N^2} such matrices. Writing an algorithm to produce such a random matrix is very easy—choose N^2 uniform random numbers $U(i, j)$, $1 \leq i, j \leq N$, and set

$$M(i, j) = \begin{cases} 0, & \text{if } U(i, j) < .5 \\ 1, & \text{if } U(i, j) \geq .5 \end{cases}.$$

It takes on the order of N^2 operations to produce one $N \times N$ matrix, and clearly every matrix in S has the same chance of being produced.

Now suppose we change our probability space and say we are only interested in matrices in S that have no two 1s together. Let T be the matrices in S with no two 1s together, i.e., the matrices $M \in S$ such that

$$M(i - 1, j) = M(i + 1, j) = M(i, j - 1) = M(i, j + 1) = 0,$$

if $M(i, j) = 1$. Suppose also we want to put the uniform probability measure on T (this is a natural measure from the perspective of statistical physics where 1s can denote particles and there is a repulsive interaction that keeps particles from getting too close together). While it is easy to define this measure, it is a hard problem to determine $c(N)$, the number of elements of T. It can be shown that there is a constant $\beta \in (1,2)$ such that

$$\lim_{N \to \infty} c(N)^{1/N^2} = \beta$$

(so that the number of elements in T is approximately β^{N^2}) but the exact value of β is not known. Still we might be interested in the properties of such matrices and hence would like to sample from the uniform distribution on T.

While it is very difficult to give an efficient algorithm that exactly samples from the uniform distribution (and even if we had one, the errors in the random number generation would keep it from being an exact sampling), we can give a very efficient algorithm that produces samples from an almost uniform distribution. What we do is run an irreducible Markov chain with state space T whose invariant measure is the uniform distribution. We can then start with any matrix in T; run the chain long enough so that the chain is near equilibrium; and then choose the matrix we have at that point.

For this example, the algorithm is as follows: 1) start with any matrix $M \in T$, e.g., the matrix with all zero entries; 2) choose one of the entries at random, i.e., choose an ordered pair (i,j) from the uniform distribution on the N^2 ordered pairs; and 3) consider the matrix gotten by changing only the (i,j) entry of M. If this new matrix is in T, we let this be the new value of the chain; if the new matrix is not in T, we make no change in the value of the chain; return to 2). This algorithm is a simulation of the discrete-time Markov chain with state space T and transition probabilities

$$P(M, M') = N^{-2},$$

if M, $M' \in T$ differ in exactly one entry; $P(M, M') = 0$ if M and M' differ by more than one entry; and $P(M, M)$ is whatever is necessary so that the rows add up to 1. Clearly, P is a symmetric matrix and it is not too difficult to see that it is irreducible. Hence P is a reversible Markov chain with state space T and invariant distribution of the uniform measure.

Of course, we need to know how long to run the chain in order to guarantee that one is close to the invariant distribution. As noted in the previous section, this boils down to estimating the second eigenvalue for the Markov chain. Unfortunately, estimating this eigenvalue is often much more difficult than showing that the chain has the right invariant measure (which is quite easy in this example). In this example, we clearly need at least N^2 steps to get close, since each of the entries should have a good chance to be changed. It can be shown, however, that we are fairly close after order N^2 steps, and if we allow about $N^2 \log N$ steps we can be very close to the invariant measure.

We will give some other examples of where these kinds of algorithms have been used. In all of these cases the algorithms are fairly efficient, although in some cases only partial rigorous analysis has been given.

Example 1. Ising Model. Let S be the set of $N \times N$ matrices with entries 1 or -1. For any $M \in S$ we define the "energy" of the matrix by

$$H(M) = - \sum_{(i,j)\sim(i',j')} M(i,j)M(i',j'),$$

where $(i,j) \sim (i',j')$ if the entries are "nearest neighbors,"

$$|i - i'| + |j - j'| = 1.$$

The value $M(i,j)$ is called the "spin" at site (i,j) and the energy is minimized when all the spins are the same. The Ising model gives a probability distribution on S that weights matrices of low energy the highest. For any $a > 0$ we let

$$\pi_a(M) = \frac{\exp\{-aH(M)\}}{\sum_{M'\in S}\exp\{-aH(M')\}}.$$

This is a well-defined probability measure, although it is difficult to calculate the normalization factor

$$Z(a) = \sum_{M'\in S} \exp\{-aH(M')\}.$$

If M and M' are two matrices that agree in all but one entry, we can calculate $\pi_a(M)/\pi_a(M')$ easily without calculating $Z(a)$.

Write $M \sim M'$ if M and M' differ in exactly one entry. We define P_a by

$$P_a(M, M') = \frac{1}{N^2} \min\{1, \frac{\pi_a(M')}{\pi_a(M)}\}, \quad M \sim M'$$

and

$$P_a(M, M) = 1 - \frac{1}{N^2} \sum_{M'\sim M} P_a(M, M').$$

In other words, one runs an algorithm as follows: 1) start with a matrix M; 2) choose an entry of the matrix at random and let M' be the matrix which agrees with M everywhere except at that entry; 3) move to matrix M' with probability $\min\{1, \pi_a(M')/\pi_a(M)\}$ and otherwise stay at the matrix M. It is easy to check that this is an irreducible Markov chain reversible with respect to π_a.

Example 2. The above example can be generalized. Suppose $G = (V, E)$ is a connected graph such that each vertex is adjacent to at most K other vertices. Suppose a positive function f on V is given, and let π be the probability measure

$$\pi(v) = \frac{f(v)}{\sum_{w\in V} f(w)}.$$

Write $v \sim w$ if $(v, w) \in E$ and set

$$P(v, w) = \frac{1}{K} \min\{1, \frac{f(w)}{f(v)}\},$$

and

$$P(v,v) = 1 - \sum_{w \sim v} P(v,w).$$

Then P is an irreducible Markov chain, reversible with respect to π. Algorithms of this type are often referred to as Metropolis algorithms.

Example 3. There is another class of algorithms, called Gibbs samplers, which are similar. Suppose we have n variables $(x_1, \ldots, x_n)$ each of which can take on one of K values say $\{a_1, \ldots, a_K\}$. Let S be the set of K^n possible n-tuples and assume we have a positive function f on S. We want to sample from the distribution

$$\pi(x_1, \ldots, x_n) = \frac{f(x_1, \ldots, x_n)}{\sum_{(y_1,\ldots,y_n) \in S} f(y_1, \ldots, y_n)}. \tag{7.2}$$

Our algorithm is to choose a $j \in \{1, \ldots, n\}$ at random and then change x_j to z according to the conditional probability

$$\frac{f(x_1, \ldots, x_{j-1}, z, x_{j+1}, \ldots, x_n)}{\sum_{k=1}^{K} f(x_1, \ldots, x_{j-1}, a_k, x_{j+1}, \ldots, x_n)}.$$

This gives the transition probability

$$P((x_1, \ldots, x_n), (y_1, \ldots, y_n)) =$$

$$\frac{1}{n} \frac{f(x_1, \ldots, x_{j-1}, y_j, x_{j+1}, \ldots, x_n)}{\sum_{k=1}^{K} f(x_1, \ldots, x_{j-1}, a_k, x_{j+1}, \ldots, x_n)}, \quad y_j \neq x_j; \ y_i = x_i, i \neq j,$$

and $P((x_1, \ldots, x_n), (x_1, \ldots, x_n))$ equal to whatever is necessary to make the rows sum to 1 Again it is straightforward to check that this is an irreducible Markov chain, reversible with respect to π. Note also that to run the chain we never need to calculate the denominator in (7.2).

The Ising model can be considered one example with $n = N^2$, $K = 2$, and the possible values $-1, 1$. In this case we get

$$P(M, M') = \frac{1}{N^2} \frac{\exp\{-aH(M)'\}}{\exp\{-aH(M)\} + \exp\{-aH(M')\}},$$

if M and M' differ in exactly one entry.

7.4 A Criterion for Recurrence

In this section we develop a useful monotonicity result for random walks with symmetric rates. To illustrate the usefulness of the result consider two possible rates on Z^2. The first is $\alpha(x,y) = 1$ if $|x - y| = 1$ and 0 otherwise. This corresponds to simple random walk which we have already seen is recurrent in two dimensions. For the other rate, suppose we remove some edges from the integer lattice—more precisely, suppose we have a subset B

of the edges of the lattice and state that $\alpha(x,y) = 1$ only if the edge (x,y) is contained in B.

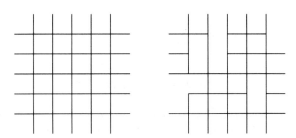

What our result will say is that for *any* such subset B the corresponding chain is still recurrent. Assume we have a graph $G = (V, E)$ and two symmetric rate functions α and β on E.

Fact. If α produces a recurrent chain and $\beta(x,y) \le \alpha(x,y)$ for all (x,y), then β also produces a recurrent chain.

The proof of this statement takes a little work. We start with some preliminary remarks. Suppose we write the elements of V as $\{x_0, x_1, x_2, \ldots\}$ (we will assume V is infinite, for otherwise the chains are always recurrent). Let $A_n = \{x_0, x_n, x_{n+1}, \ldots\}$. Let us start the chain at x_0, wait until it leaves x_0 for the first time, and then see what point in A_n is hit first by the chain. Let $h_n(x_0) = h_n(x_0; \alpha)$ be the probability that the first such point hit is not x_0 (using transition rates α). Then it is not too difficult to convince oneself that the chain is recurrent if and only if

$$\lim_{n \to \infty} h_n(x_0) = 0. \tag{7.3}$$

It is the goal of this section to give a formulation of $h_n(x_0)$ that will allow us to conclude the monotonicity result.

For this section we will assume that a graph $G = (V, E)$ is given as well as a symmetric transition rate $\alpha : E \to [0, \infty)$. Let A be a subset of V and fix $x_0 \in A$. Let X_t be a continuous time Markov chain with rates α and let τ be the infimum of all $t \ge 0$ such that $X_t \in A$. Define $f(y)$ to be the probability starting at y that the first visit to A occurs at the point x_0,

$$f(y) = P\{X_\tau = x_0 \mid X_0 = y\}.$$

It is easy to see that $f(x_0) = 1$ and $f(y) = 0$ for $y \in A$, $y \neq x_0$. Suppose $y \notin A$. Then the probability that the first new site that y visits is z is $\alpha(y, z)/\alpha(y)$, where again we write $\alpha(y) = \sum_{z \in V} \alpha(y, z)$. By concentrating on this first move, we see that

$$
\begin{aligned}
f(y) &= \sum_{z \in V} P\{\text{first new site is } z\} f(z) \\
&= \sum_{z \in V} \frac{\alpha(y, z)}{\alpha(y)} f(z),
\end{aligned}
$$

or

$$
\alpha(y) f(y) = \sum_{z \in V} \alpha(y, z) f(z). \tag{7.4}
$$

A function f satisfying (7.4) is called α-harmonic at y. We have shown that our given f is α-harmonic at all $y \notin A$, and one can show with a little more work that f is the unique function that is α-harmonic at $y \notin A$ and that satisfies the boundary condition $f(x_0) = 1$, $f(y) = 0$, $y \in A, y \neq x$.

We will now characterize f as the function that minimizes a particular functional (a functional is a function of a function). For any function g let

$$
Q_\alpha(g) = \sum_{x \in V} \sum_{y \in V} \alpha(x, y)(g(x) - g(y))^2.
$$

Suppose we consider only those functions g that satisfy the boundary condition $g(x_0) = 1$, $g(y) = 0$, $y \in A, y \neq x$. Let $\bar{g}$ be the function satisfying this boundary condition which minimizes Q_α. Then at any $y \notin A$, perturbations of $\bar{g}$ at y, leaving all other values fixed, should increase Q_α. In other words if we define $\bar{g}_\epsilon(z)$ by

$$
\bar{g}_\epsilon(z) = \begin{cases} \bar{g}(z), & z \neq y, \\ \bar{g}(y) + \epsilon, & z = y, \end{cases}
$$

Then

$$
\frac{d}{d\epsilon} Q_\alpha(\bar{g}_\epsilon) \big|_{\epsilon=0} = 0.
$$

A simple calculation shows that this holds if and only if for every $y \notin A$,

$$
\sum_{z \in V} \bar{g}(z) \alpha(y, z) = \sum_{z \in V} \bar{g}(y) \alpha(y, z) = \bar{g}(y) \alpha(y).
$$

In other words $\bar{g}$ is the function that is α-harmonic at each $y \notin A$ and satisfies the boundary conditions. Since f is the only such function, $\bar{g} = f$. Summarizing, f, as defined above, is also the function that minimizes $Q_\alpha(g)$ subject to the boundary condition, $g(x_0) = 1$, $g(y) = 0$, $y \in A, y \neq x$.

We now use "summation by parts" to give another expression for $Q_\alpha(f)$. We start by writing

$$
Q_\alpha(f) = \sum_{x \in V} \sum_{y \in V} \alpha(x, y)(f(x) - f(y))^2
$$

$$= \sum_x \sum_y \alpha(x,y)f(x)(f(x) - f(y))$$

$$- \sum_x \sum_y \alpha(x,y)f(y)(f(x) - f(y))$$

$$= 2 \sum_x \sum_y \alpha(x,y)f(x)(f(x) - f(y)).$$

The last equality uses the symmetry of α. Since $f(x_0) = 1$ and $f(y) = 0$, $y \in A, y \neq x$ we can write this as

$$2 \sum_y \alpha(x_0, y)(1 - f(y)) + 2 \sum_{x \notin A} f(x) \sum_y \alpha(x,y)(f(x) - f(y)).$$

But, if $x \notin A$, then f is α-harmonic at x,

$$\sum_y \alpha(x,y)f(y) = \sum_y \alpha(x,y)f(x) = \alpha(x)f(x).$$

Hence the second term in the sum is 0 and we get

$$Q_\alpha(f) = 2 \sum_{y \in V} \alpha(x_0, y)(1 - f(y)) = 2\alpha(x_0) \sum_{y \in V} \frac{\alpha(x_0, y)}{\alpha(x_0)} (1 - f(y)).$$

Now let $h(x_0)$ be the probability that the chain starting at x_0 makes its first visit to A, after leaving x_0 for the first time, at some point other than x_0. By considering the first step, we see that

$$h(x_0) = \sum_{y \in V} \frac{\alpha(x_0, y)}{\alpha(x_0)} (1 - f(y)) = \frac{Q_\alpha(x_0, A)}{2\alpha(x_0)},$$

where

$$Q_\alpha(x_0, A) = \inf \sum_{x \in V} \sum_{y \in V} \alpha(x,y)(g(x) - g(y))^2,$$

and the infimum is taken over all functions g satisfying $g(x_0) = 0$ and $g(y) = 1$, $y \in A, y \neq x$. The beauty of the formula comes in the realization that if $\beta(x,y)$ is another collection of rates with $\beta(x,y) \leq \alpha(x,y)$ for all $x, y \in V$, then

$$Q_\beta(x_0, A) \leq Q_\alpha(x_0, A).$$

If we now write $h_n(x_0; \alpha)$ and $h_n(x_0, \beta)$ as in the beginning of this section, we see that we have shown that if $\beta(x,y) \leq \alpha(x,y)$ for all x, y,

$$h_n(x_0; \beta) \leq \frac{\beta(x_0)}{\alpha(x_0)} h_n(x_0; \alpha).$$

In particular, if we use the criterion given in (7.3), we see that if the chain with rates α is recurrent, then the chain with rates β is also recurrent.

7.5 Exercises

7.1 Show that every irreducible, discrete-time, two-state Markov chain is reversible with respect to its invariant probability.

7.2 COMPUTER SIMULATION. Let M be a matrix chosen uniformly from the set of 50×50 matrices with entries 0 and 1 such that no two 1s are together (see Section 7.3). Use a Markov chain simulation as described in Section 7.3 to estimate the probability that the $M(25, 25)$ entry of this matrix is a 1.

7.3 Find the eigenvalues of the $N \times N$ matrix $\mathbf{A}$ from Example 2, Section 7.2,

$$A(i, j) = \begin{cases} -1, & i = j, \\ 1/2, & |i - j| = 1 (mod N), \\ 0, & \text{otherwise.} \end{cases}$$

[Hint: any eigenvector with eigenvalue λ can be considered as a function $f(n)$ on the integers satisfying

$$\lambda f(n) = \frac{1}{2} f(n + 1) + \frac{1}{2} f(n - 1) - f(n),$$

$$f(n) = f(n + N),$$

for each n. Find the general solution of the difference equation and then use the periodicity condition to put restrictions on the λ.]

7.4 Let $\alpha(x, y)$ be a symmetric rate function on the edges of the integer lattice Z^d, i.e., a nonnegative function defined for all $x, y \in Z^d$ with $|x - y| = 1$ that satisfies $\alpha(x, y) = \alpha(y, x)$. Suppose there exist numbers $0 < c_1 < c_2 < \infty$ such that for all x, y with $|x - y| = 1$,

$$c_1 \leq \alpha(x, y) \leq c_2.$$

Let X_t be a continuous-time Markov chain with rates $\alpha(x, y)$.
  (a) If $d = 1, 2$, show that the chain is recurrent.
  (b) If $d \geq 3$, show that the chain is transient.

CHAPTER 8

Brownian Motion

8.1 Introduction

Brownian motion is a stochastic process that models random continuous motion. In order to model "random continuous motion," we start by writing down the physical assumptions that we will make. Let X_t represent the position of a particle at time t. In this case t takes on values in the non-negative real numbers and X_t takes on values in the real line (or perhaps the plane or space). This will be an example of a stochastic process with both continuous time and continuous state space.

For ease we will start with the assumption $X_0 = 0$. The next assumption is that the motion is "completely random." Consider two times $s < t$. We do not wish to say that the positions X_s and X_t are independent, but rather that the motion after time s, $X_t - X_s$, is independent of X_s. We will need this assumption for any finite number of times: for any $s_1 \le t_1 \le s_2 \le t_2 \le \cdots \le s_n \le t_n$, the random variables $X_{t_1} - X_{s_1}, X_{t_2} - X_{s_2}, \ldots, X_{t_n} - X_{s_n}$ are independent. Also the distribution of the random movements should not change with time. Hence we will assume that the distribution of $X_t - X_s$ depends only on $t - s$. For the time being, we will also assume that there is no "drift" to the process, i.e., $E(X_t) = 0$.

The above assumptions are not sufficient to describe the model we want. In fact, if Y_t is the Poisson process and $X_t = Y_t - t$ [so that $E(X_t) = 0$], X_t satisfies these assumptions but is clearly not a model for continuous motion. We will include as our final assumption for our model this continuity: the function X_t is a continuous function of t.

It turns out that the above assumptions uniquely describe the process at least up to a scaling constant. Suppose the process X_t satisfies these assumptions. What is the distribution of the random variable X_t? For ease, we will discuss the case $t = 1$. For any n, we can then write

$$X_1 = [X_{1/n} - X_0] + [X_{2/n} - X_{1/n}] + \cdots + [X_{n/n} - X_{(n-1)/n}].$$

In other words, X_1 can be written as the sum of n independent, identically distributed random variables. Moreover, if n is large, each of the random variables is small. To be more precise, if we let

$$M_n = \max\{|X_{1/n} - X_0|, |X_{2/n} - X_{1/n}|, \ldots |X_{n/n} - X_{(n-1)/n}|\},$$

then as $n \to \infty$, $M_n \to 0$. This is a consequence of the assumption that X_t is a continuous function of t (if M_n did not go to 0 then there would be a "jump" in the path of X_t). It is a theorem of probability theory that the only distribution that can be written as the sum of n independent, identically distributed random variables such that the maximum of the variables goes to 0 is a normal distribution. We can thus conclude that the distribution of X_t is a normal distribution. We now formalize this definition.

Definition. A *Brownian motion* or a *Wiener process with variance parameter* σ^2 is a stochastic process X_t taking values in the real numbers satisfying

(i) $X_0 = 0$;

(ii) For any $s_1 \le t_1 \le s_2 \le t_2 \le \cdots \le s_n \le t_n$, the random variables $X_{t_1} - X_{s_1}, \ldots, X_{t_n} - X_{s_n}$ are independent;

(iii) For any $s < t$, the random variable $X_t - X_s$ has a normal distribution with mean 0 and variance $(t - s)\sigma^2$;

(iv) The paths are continuous, i.e., the function $t \longmapsto X_t$ is a continuous function of t.

While it is standard to include the fact that the increments are normally distributed in the definition, it is worth remembering that this fact can actually be deduced from the physical assumptions. *Standard Brownian motion* is a Brownian motion with $\sigma^2 = 1$. We can also speak of a Brownian motion starting at x; this is a process satisfying conditions (ii) – (iv) and the initial condition $X_0 = x$. If X_t is a Brownian motion (starting at 0), then $Y_t = X_t + x$ is a Brownian motion starting at x.

Brownian motion can be constructed as a limit of random walks. Suppose S_n is an unbiased random walk on the integers. We can write

$$S_n = Y_1 + \cdots + Y_n,$$

where the random variables Y_i are independent,

$$P\{Y_i = 1\} = P\{Y_i = -1\} = 1/2.$$

Now instead of having time increments of size 1 we will have increments of size $\Delta t = 1/N$ where N is an integer. We will set

$$W_{k\Delta t}^{(N)} = a_N S_k,$$

where we choose a normalizing constant a_N so that W_1 has variance 1. Since $Var(S_N) = N$, it is clear that we must choose $a_N = N^{-1/2}$. Hence in this discrete approximation, the size of the jump in time $\Delta t = 1/N$ is $1/\sqrt{N} = (\Delta t)^{1/2}$. We can consider the discrete approximation as a process for all values of t by linear interpolation (see the figure below).

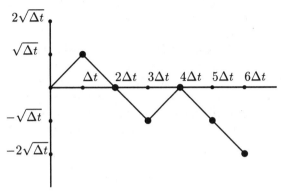

As $N \to \infty$, this discrete approximation approaches a continuous-time, continuous-space process. By the central limit theorem the distribution of

$$W_1^{(N)} = N^{-1/2} S_N$$

approaches a normal distribution with mean 0 and variance 1. Similarly, the distribution of $W_t^{(N)}$ approaches a normal distribution with mean 0 and variance t. The limiting process can be shown to be a standard Brownian motion. (It requires some sophisticated mathematics to state explicitly what kind of limit is being taken here. We will not worry about this detail.)

The path of a Brownian motion is very rough. Consider the increment $X_{t+\Delta t} - X_t$ for small Δt. The distribution of this increment has mean 0, variance Δt so

$$E(|X_{t+\Delta t} - X_t|^2) = \Delta t.$$

In other words the typical size of an increment, $|X_{t+\Delta t} - X_t|$, is about $\sqrt{\Delta t}$. As $\Delta t \to 0$, $\sqrt{\Delta t} \to 0$, which is consistent with the continuity of the paths. What about differentiability? Does it make sense to talk about dX_t/dt? Recall the definition of the derivative,

$$\frac{dX_t}{dt} = \lim_{\Delta t \to 0} \frac{X_{t+\Delta t} - X_t}{\Delta t}.$$

When Δt is small, the absolute value of the numerator is on the order of $\sqrt{\Delta t}$ which is much larger than Δt. Hence, this limit does not exist. By a sharpening of this argument one can prove the following.

Fact. The path of a Brownian motion X_t is nowhere differentiable.

Care is needed in proving statements such as the one above. The intuitive argument can be used fairly easily to prove the statement "for each t, the probability that X_t is not differentiable at t is 1." This is not as strong as the fact above which states "the probability that X_t is not differentiable at all values of t is 1." This is a little tricky to understand. As a possibly easier example consider the following two statements: "For each t, the probability that $X_t \neq 1$ is 1" and "The probability that $X_t \neq 1$ for all values of t is 1." These statements are not the same, and, in fact, the first is true and the second is false. For any given t, X_t has a normal distribution; hence the probability of taking on any particular value is 0 (this is true for any continuous distribution). However, the probability that $X_1 > 1$ is certainly greater than 0. If $X_0 = 0$ and $X_1 > 1$, then the continuity of X_t implies that $X_t = 1$ for some $0 < t < 1$. Hence the probability that $X_t = 1$ for some $0 < t < 1$ is greater than 0. The difficulty here comes with the fact that the real numbers are uncountable. We can write

$$\{X_t = 1 \text{ for some } 0 \leq t \leq 1\} = \cup_{0 \leq t \leq 1}\{X_t = 1\}.$$

The right-hand side is a union of sets each with probability 0. However, it is an *uncountable* union of such sets. The axioms of probability imply that the countable union of sets of probability 0 has probability 0 but does not say the same for an uncountable union. This phenomenon arises whenever one deals in continuous probability. For example, if Y is any continuous random variable then

$$\{-\infty < Y < \infty\} = \cup_{-\infty < y < \infty}\{Y = y\}.$$

The right-hand side is a union of events with probability 0, but the left-hand side has probability 1.

In stochastic processes with continuous time and space, many difficult technical problems can arise in trying to deal with uncountable unions of sets. We will ignore most of these issues here. (Most of these problems are relatively easily overcome for Brownian motion.)

8.2 Markov Property

Let X_t be a standard Brownian motion. We will let $\mathcal{F}_t$ represent the information contained in $X_s, s \leq t$, in other words all the information that can be obtained from watching the Brownian motion up through time t. Suppose $s < t$ and consider the conditional expectation $E(X_t \mid \mathcal{F}_s)$. Note that

$$E(X_t \mid \mathcal{F}_s) = E(X_s \mid \mathcal{F}_s) + E(X_t - X_s \mid \mathcal{F}_s).$$

Since X_s is $\mathcal{F}_s$ measurable, the first term on the right-hand side equals X_s. Since $X_t - X_s$ is independent of $\mathcal{F}_s$, the second term equals $E(X_t - X_s) = 0$. Hence

$$E(X_t \mid \mathcal{F}_s) = X_s = E(X_t \mid X_s).$$

This equation illustrates the Markov property of Brownian motion, i.e., in order to predict X_t given all the information up through time s, it suffices to consider only the value of the Brownian motion at time s.

Let $p_t(x, y)$ denote the transition densities, i.e., the density of X_t for Brownian motion starting at x. Since $X_t - X_0$ is normal, mean 0, variance t,

$$p_t(x, y) = \frac{1}{\sqrt{2\pi t}} e^{-(y-x)^2/2t}, \quad -\infty < y < \infty.$$

The transition densities satisfy the *Chapman–Kolmogorov equation*

$$p_{s+t}(x, y) = \int_{-\infty}^{\infty} p_s(x, z) p_t(z, y) \, dz.$$

This can be verified directly for this transition function, but one can also see this by appealing to the Markov property (the Chapman–Kolmogorov equation holds in general for Markov processes). This property states that $Y_t = X_{s+t} - X_s$ is a Brownian motion independent of X_s; in other words $Z_t = X_{s+t}$ is a Brownian motion starting at the (random) starting point X_s. The Chapman–Kolmogorov equation then just averages the density $p_t(z, y)$ over all possible starting points z.

In order to do many useful computations about Brownian motions, a more general Markov property is needed. This is generally referred to as the *strong Markov property*. We first need the notion of a real valued stopping time. The definition is a generalization of the definition of a stopping time given for discrete-time processes. We say that a random variable T taking values in $[0, \infty]$ is a *stopping time* for Brownian motion if for each t the (indicator function of the) event $\{T \leq t\}$ is measurable with respect to $\mathcal{F}_t$. In other words, to know whether or not the process has stopped before time t, one only needs to look at the Brownian motion up through time t. The most important examples will be stopping times of the form

$$T_x = \inf\{t : X_t = x\}.$$

If T is a stopping time, we write $\mathcal{F}_T$ for the information contained in the Brownian motion up through the stopping time T (one gets to view the path up through time T but not beyond). We will let Y_t denote the process beyond time T,

$$Y_t = X_{t+T} - X_T.$$

Then the strong Markov property states that Y_t is a Brownian motion independent of $\mathcal{F}_T$.

It is easier to see what this means by considering an example of how the property is used. Suppose the Brownian motion starts at 0 and we want to calculate the probability that there exists some t with $0 \leq t \leq 1$ and $X_t \geq 1$. Let $T = T_1$ be the first time that the Brownian motion equals 1. Then, by continuity, the event $\{X_t \geq 1 \text{ for some } 0 \leq t \leq 1\}$ is the same as the event $\{T \leq 1\}$. Since

$$P\{T = 1\} \leq P\{X_1 = 1\} = 0,$$

we can see that

$$P\{T \le 1\} = P\{T < 1\}.$$

Now consider the event $\{X_1 \ge 1\}$. Since X_1 is normal, mean 0, variance 1,

$$P\{X_1 \ge 1\} = \int_1^\infty \frac{1}{\sqrt{2\pi}} e^{-x^2/2} \, dx.$$

Also,

$$P\{X_1 \ge 1\} = P\{T \le 1\} P\{X_1 \ge 1 \mid T \le 1\}.$$

Now we use the strong Markov property. Suppose $T \le 1$. We may assume in fact that $T < 1$ (since $T = 1$ has probability 0 of occurring). Then, given T, $X_1 - X_T = X_1 - 1$ is a normal random variable, mean 0, variance $1 - T$. Regardless of the variance, we know by the symmetry of the normal distribution that the probability that this normal random variable is greater than or equal to 0 is $1/2$. Hence, we conclude

$$P\{X_1 - 1 \ge 0 \mid T \le 1\} = 1/2.$$

Therefore

$$P\{T \le 1\} = 2P\{X_1 \ge 1\} = 2 \int_1^\infty \frac{1}{\sqrt{2\pi}} e^{-x^2/2} \, dx.$$

This result is a particular case of the reflection principle. We now state the general result which is proved in the same way.

Reflection Principle. Suppose X_t is a Brownian motion with variance parameter σ^2 starting at a and $a < b$. Then for any $t > 0$,

$$\begin{aligned}
P\{X_s \ge b \text{ for some } 0 \le s \le t\} &= 2P\{X_t \ge b\} \\
&= 2 \int_b^\infty \frac{1}{\sqrt{2\pi t \sigma^2}} e^{-(x-a)^2/2\sigma^2 t} \, dx.
\end{aligned}$$

Example 1. Let $t > 1$ and let us compute the probability that a standard Brownian motion crosses the x-axis sometime between times 1 and t, i.e.,

$$P\{X_s = 0 \text{ for some } 1 \le s \le t\}.$$

We first condition on what happens at time $t = 1$. Suppose $X_1 = b > 0$. Then the probability that $X_s = 0$ for some $1 \le s \le t$ is the same as the probability that $X_s \le -b$ for some $0 \le s \le t - 1$. This is the same (by symmetry) as the probability that $X_s \ge b$ for some $0 \le s \le t - 1$. This probability is given by the reflection principle, so

$$P\{X_s = 0 \text{ for some } 1 \le s \le t \mid X_1 = b\}$$

$$= 2 \int_b^\infty \frac{1}{\sqrt{2\pi(t-1)}} e^{-x^2/2(t-1)} \, dx.$$

By symmetry, again, the probability is the same if $X_1 = -b$. Hence, by averaging over all possible values of b we get

$$P\{X_s = 0 \text{ for some } 1 \leq s \leq t\}$$

$$= \int_{-\infty}^{\infty} p_1(0, b) P\{X_s = 0 \text{ for some } 1 \leq s \leq t \mid X_1 = b\}\, db$$

$$= 2 \int_0^{\infty} \frac{1}{\sqrt{2\pi}} e^{-b^2/2} \left[2 \int_b^{\infty} \frac{1}{\sqrt{2\pi(t-1)}} e^{-x^2/2(t-1)}\, dx\right] db.$$

The substitution $y = x/\sqrt{t-1}$ in the inside integral reduces this integral to

$$4 \int_0^{\infty} \int_{b(t-1)^{-1/2}}^{\infty} \frac{1}{2\pi} e^{-(b^2+y^2)/2}\, dy\, db.$$

This integral can be computed using polar coordinates. Note that the region $\{0 < b < \infty, b(t-1)^{-1/2} < y < \infty\}$ corresponds to the polar region $\{0 < r < \infty, \arctan(\sqrt{t-1})^{-1} < \theta < \pi/2\}$. Hence the probability equals

$$4 \int_0^{\infty} \int_{\arctan((\sqrt{t-1})^{-1})}^{\pi/2} \frac{1}{2\pi} e^{-r^2/2}\, r\, d\theta\, dr$$

$$= 4\left(\frac{\pi}{2} - \arctan \frac{1}{\sqrt{t-1}}\right) \frac{1}{2\pi} \int_0^{\infty} r e^{-r^2/2} dr$$

$$= 1 - \frac{2}{\pi} \arctan \frac{1}{\sqrt{t-1}}.$$

Example 2. We will show that (with probability 1)

$$\lim_{t \to \infty} t^{-1} X_t = 0.$$

First, we consider the limit taken over only integer times. Note that for n an integer,

$$X_n = (X_1 - X_0) + \ldots + (X_n - X_{n-1}),$$

is a sum of independent, identically distributed random variables. It follows from the (strong) law of large numbers that

$$\lim_{n \to \infty} n^{-1} X_n = 0.$$

For each n, let

$$M_n = \sup\{|X_t - X_n| : n \leq t \leq n+1\}.$$

If we can show that

$$\lim_{n \to \infty} n^{-1} M_n = 0,$$

we will be finished since for any t, if $n \leq t < n+1$,

$$t^{-1}|X_t| \leq n^{-1}|X_t| \leq n^{-1}(|X_n| + |M_n|).$$

For any $a > 0$, symmetry and the reflection principle state that

$$P\{|M_n| \geq a\} \leq 2P\{M_n \geq a\} \quad = \quad 4\int_a^\infty \frac{1}{\sqrt{2\pi}} e^{-x^2/2} \, dx$$

$$\leq \quad 4\int_a^\infty \frac{1}{\sqrt{2\pi}} e^{-xa/2} \, dx$$

$$= \quad \frac{8}{a\sqrt{2\pi}} e^{-a^2/2}.$$

If we plug in $a = 2(\ln n)^{1/2}$, we get

$$P\{|M_n| \geq 2\sqrt{\ln n}\} \leq \frac{8}{2\sqrt{2\pi \ln n}\, n^2}.$$

In particular, for all n sufficiently large, the probability is less than n^{-2}. If we let I_n denote the indicator function of the event $\{|M_n| \geq 2\sqrt{\ln n}\}$ and

$$I = \sum_{n=0}^\infty I_n,$$

we find that $E(I) < \infty$. This states that the expected number of times that $|M_n| \geq 2\sqrt{\ln n}$ is finite and hence that, with probability 1, $|M_n| \geq 2\sqrt{\ln n}$ only finitely often. In particular, this implies that $n^{-1}M_n \to 0$.

8.3 Zero Set of Brownian Motion

In this section we will investigate the (random) set

$$Z = \{t : X_t = 0\}.$$

It turns out that this set is an interesting "fractal" subset of the real line.

In analyzing this set we will use two scaling results about Brownian motion which will be proved in the exercises (see Exercises 8.6 and 8.7). Suppose X_t is a standard Brownian motion. Then,

(1) If $a > 0$, and $Y_t = a^{-1/2} X_{at}$, then Y_t is a standard Brownian motion.

(2) If X_t is a standard Brownian motion and $Y_t = tX_{1/t}$, then Y_t is a standard Brownian motion.

In an example in the previous section, we proved that

$$P\{Z \cap [1, t] \neq \emptyset\} = 1 - \frac{2}{\pi} \arctan \frac{1}{\sqrt{t-1}}.$$

As $t \to \infty$ the quantity on the right-hand side tends to 1. This tells us that with probability 1 the Brownian motion eventually returns to the origin, and hence (with the help of the strong Markov property) that it returns infinitely often. This means that the Brownian motion for large t has both positive and negative values.

What happens near $t = 0$? Let $Y_t = tX_{1/t}$. Then Y_t is also a standard Brownian motion. As time goes to infinity in the process X, time goes to 0 in Y. Hence, since X_t has both positive and negative values for arbitrarily large values of t, Y_t has positive and negative values for arbitrarily small values of t. This states that in any interval about 0 the Brownian motion takes on both positive and negative values (and hence by continuity also the value 0)!

One topological property that Z satisfies is the fact that Z is a closed set. This means that if a sequence of points $t_i \in Z$ and $t_i \to t$, then $t \in Z$. This follows from the continuity of the function X_t. For any continuous function, if $t_i \to t$, then $X_{t_i} \to X_t$. We have seen that 0 is not an isolated point of Z, i.e., there are positive numbers $t_i \in Z$ such that $t_i \to 0$. It can be shown that none of the points of Z are isolated points. From a topological perspective Z looks like the Cantor set (see the example below for a definition).

How "big" is the set Z? To discuss this we need to discuss the notion of a dimension of a set. There are two similar notions of dimension, Hausdorff dimension and box dimension, which can give fractional dimensions to sets. (There is a phrase "fractal dimension" which is used a lot in scientific literature. As a rule, the people who use this phrase are not distinguishing between Hausdorff and box dimension and could mean either one.) The notion of dimension we will discuss here will be that of box dimension, but all the sets we will discuss have Hausdorff dimension equal to their box dimension. Suppose we have a bounded set A in d-dimensional space R^d. Suppose we cover A with d-dimensional balls of diameter ϵ. How many such balls are needed? If A is a line segment of length 1 (one-dimensional set), then ϵ^{-1} such balls are needed. If A is a two-dimensional square, however, on the order of ϵ^{-2} such balls are needed. One can see that for a standard k-dimensional set, we need ϵ^{-k} such balls. This leads us to define the (box) dimension of the set A to be the number D such that for small ϵ the number of balls of diameter ϵ needed to cover A is on the order of ϵ^{-D}.

Example. Consider the fractal subset of $[0, 1]$, the Cantor set. The Cantor set A can be defined as a limit of approximate Cantor sets A_n. We start with $A_0 = [0, 1]$. The next set A_1 is obtained by removing the open middle interval $(1/3, 2/3)$, so that

$$A_1 = [0, \frac{1}{3}] \cup [\frac{2}{3}, 1].$$

The second set A_2 is obtained by removing the middle thirds of the two intervals in A_1, hence

$$A_2 = [0, \frac{1}{9}] \cup [\frac{2}{9}, \frac{1}{3}] \cup [\frac{2}{3}, \frac{7}{9}] \cup [\frac{8}{9}, 1].$$

In general A_{n+1} is obtained from A_n by removing the "middle third" of each interval. The Cantor set A is then the limit of these sets A_n,

$$A = \cap_{n=1}^{\infty} A_n.$$

Note that A_n consists of 2^n intervals each of length 3^{-n}. Suppose we try to cover A by intervals of length 3^{-n},

$$[\frac{k-1}{3^n}, \frac{k}{3^n}].$$

We need 2^n such intervals. Hence the dimension D of the Cantor set is the number such that $2^n = (3^{-n})^{-D}$, i.e.,

$$D = \frac{\ln 2}{\ln 3} \approx .631.$$

Now consider the set Z and consider $Z_1 = Z \cap [0, 1]$. We will try to cover Z_1 by one-dimensional balls (i.e., intervals) of diameter (length) $\epsilon = 1/n$. For ease we will consider the n intervals

$$[\frac{k-1}{n}, \frac{k}{n}], \quad k = 1, 2, \ldots n.$$

How many of these intervals are needed to cover Z_1? Such an interval is needed if $Z_1 \cap [(k-1)/n, k/n] \neq \emptyset$. What is

$$P(k, n) = P\{Z_1 \cap [\frac{k-1}{n}, \frac{k}{n}] \neq \emptyset\}?$$

Assume $k \geq 1$ (if $k = 0$, the probability is 1 since $0 \in Z$). By the scaling property of Brownian motion, $Y_t = ((k-1)/n)^{-1/2} X_{nt/(k-1)}$ is a standard Brownian motion. Hence

$$P(k, n) = P\{Y_t = 0 \text{ for some } 1 \leq t \leq \frac{k}{k-1}\}.$$

This probability was calculated in the previous section,

$$P(k, n) = 1 - \frac{2}{\pi} \arctan \sqrt{k-1}.$$

Therefore, the expected number of the intervals needed to cover Z_1 looks like

$$\sum_{k=1}^{n} P(k, n) = \sum_{k=1}^{n} [1 - \frac{2}{\pi} \arctan \sqrt{k-1}].$$

To estimate the sum, we need to consider the Taylor series for $\arctan(1/t)$ at $t = 0$ (which requires remembering the derivative of arctan),

$$\arctan \frac{1}{t} = \frac{\pi}{2} - t + O(t^2).$$

In other words, for x large,

$$\arctan x \approx \frac{\pi}{2} - \frac{1}{x}.$$

Hence

$$\sum_{k=1}^{n} P(k,n) \approx 1 + \sum_{k=2}^{n} \frac{2}{\pi\sqrt{k-1}} \approx \frac{2}{\pi} \int_{1}^{n} (x-1)^{-1/2}\, dx \approx \frac{4}{\pi}\sqrt{n}.$$

Hence it takes on the order of $\sqrt{n}$ intervals of length $1/n$ to cover Z_1, or, in other words, the dimension of the set Z is $1/2$.

8.4 Brownian Motion in Several Dimensions

Suppose $X_t^1, \ldots, X_t^d$ are independent (one-dimensional) standard Brownian motions. We will call the vector-valued stochastic process

$$X_t = (X_t^1, \ldots, X_t^d)$$

a standard d-dimensional Brownian motion. In other words, a d-dimensional Brownian motion is a process in which each component performs a Brownian motion, and the component Brownian motions are independent.

It is not difficult to show that X_t defined as above satisfies the following:
(i) $X_0 = 0$;
(ii) for any $s_1 \le t_1 \le s_2 \le t_2 \le \cdots \le s_n \le t_n$, the (vector-valued) random variables $X_{t_1} - X_{s_1}, \ldots, X_{t_n} - X_{t_{n-1}}$ are independent;
(iii) the random variable $X_t - X_s$ has a joint normal distribution with mean 0 and covariance $(t-s)I$, i.e., has density

$$f(x_1, \ldots, x_d) = \left(\frac{1}{\sqrt{2\pi r}} e^{-x_1^2/2r}\right) \cdots \left(\frac{1}{\sqrt{2\pi r}} e^{-x_d^2/2r}\right) = \frac{1}{(2\pi r)^{d/2}} e^{-|x|^2/2r},$$

where $r = t - s$;
(iv) X_t is a continuous function of t.

We could use (i) – (iv) as the definition of X_t, but we would quickly discover that we could construct X_t by taking d independent one-dimensional Brownian motions. As in the one-dimensional case we let $p_t(x,y), x,y \in R^d$ denote the probability density of X_t assuming $X_0 = x$ (it is clear how to define a Brownian motion starting at any point in R^d),

$$p_t(x,y) = \frac{1}{(2\pi t)^{d/2}} e^{-|y-x|^2/2t}.$$

Again, this satisfies the Chapman–Kolmogorov equation

$$p_{s+t}(x,y) = \int_{R^d} p_s(x,z) p_t(z,y)\, dz_1 \cdots dz_d.$$

Brownian motion is closely related to the theory of diffusion. Suppose that a large number of particles are distributed in R^d according to a density $f(y)$. Let $f(t,y)$ denote the density of the particles at time t (so that $f(0,y) = f(y)$). If we assume that the particles perform standard Brownian motions, independently, then we can write the density of particles at time t.

If a particle starts at position x, then the probability density for its position at time t is $p_t(x, y)$. By integrating, we get

$$f(t, y) = \int_{R^d} f(x) p_t(x, y) \, dx_1 \cdots dx_d.$$

The symmetry of Brownian motion tells us that $p_t(x, y) = p_t(y, x)$. Hence we can write the right-hand side as

$$\int_{R^d} f(x) p_t(y, x) \, dx_1 \cdots dx_d.$$

The right-hand side represents the expected value of $f(X_t)$ assuming $X_0 = y$. We can then write this,

$$f(t, y) = E^y(f(X_t)).$$

The notation E^y is used to denote expectations of X_t assuming $X_0 = y$.

We will now derive a differential equation that $f(t, x)$ satisfies. Consider $\partial f / \partial t$; for ease we will take $t = 0$, $d = 1$. If f is sufficiently nice, we can write the Taylor series for f about x,

$$f(y) = f(x) + f'(x)(y - x) + \frac{1}{2} f''(x)(y - x)^2 + o((y - x)^2),$$

where $o(\cdot)$ denotes an error term such that $o((y - x)^2)/(y - x)^2 \to 0$ as $y \to x$. Therefore,

$$
\begin{aligned}
\frac{\partial f}{\partial t}\Big|_{t=0} &= \lim_{t \to 0} \frac{1}{t} [E^x(f(X_t) - f(X_0))] \\
&= \lim_{t \to 0} \frac{1}{t} [f'(x) E^x [X_t - x] \\
&\quad + \frac{1}{2} f''(x) E^x [(X_t - x)^2] + o((X_t - x)^2)].
\end{aligned}
$$

We know that $E^x[X_t - x] = 0$ and $E^x[(X_t - x)^2] = Var(X_t) = t$. Also since $(X_t - x)^2$ is of order t, the term $t^{-1} o(\cdot)$ tends to 0. Hence we get

$$\frac{\partial f}{\partial t}\Big|_{t=0} = \frac{1}{2} f''(x).$$

The same argument holds for all t giving

$$\frac{\partial f}{\partial t} = \frac{1}{2} \frac{\partial^2 f}{\partial x^2}.$$

Similarly, we can extend this argument to d dimensions and show that f satisfies the equation

$$\frac{\partial f}{\partial t} = \frac{1}{2} \Delta f,$$

where Δ denotes the Laplacian,

$$\Delta f(t, x_1, \ldots, x_d) = \sum_{i=1}^{d} \frac{\partial^2 f}{\partial x_i^2}.$$

This equation is often called the *heat equation*. One can find a similar solution to the heat equation with diffusion constant D,

$$\frac{\partial f}{\partial t} = \frac{D}{2}\Delta f,$$

by considering Brownian motions with variance parameter $\sigma^2 = D$.

Sometimes it is useful to consider the heat equation in a bounded domain. Let B be a bounded region of d-dimensional space with boundary ∂B.

Imagine an initial heat distribution on B, $f(x), x \in B$ is given. Suppose also that the temperature is fixed at the boundary, i.e., there is a function $g(y), y \in \partial B$ representing the fixed temperature at point y. If $u(t, x)$ denotes the temperature at x at time t, then $u(t, x)$ satisfies

(i) $\dfrac{\partial u}{\partial t} = \dfrac{D}{2}\Delta u, \quad x \in B,$

(ii) $u(t, x) = g(x), \quad x \in \partial B,$

(iii) $u(0, x) = f(x), \quad x \in B.$

The solution of (i) – (iii) can be written in terms of Brownian motion. Let X_t be a d-dimensional Brownian motion with variance parameter $\sigma^2 = D$. Let $\tau = \tau_{\partial B}$ be the first time that the Brownian motion hits the boundary ∂B,

$$\tau = \inf\{t : X_t \in \partial B\}.$$

Then the solution can be written as

$$u(t, x) = E^x[f(X_t)I\{\tau > t\} + g(X_\tau)I\{\tau \le t\}].$$

In other words, at time t, take the average value of the following: $f(X_t)$ for the paths that have not hit ∂B and $g(X_\tau)$ for those paths that have hit ∂B. As $t \to \infty$, the temperature approaches a steady-state distribution $v(x)$ with boundary value $g(x)$. As a steady-state solution, this satisfies

(i) $\Delta v(x) = 0, \quad x \in B,$

(ii) $v(x) = g(x), \quad x \in \partial B$.

The solution is given by

$$v(x) = \lim_{t\to\infty} u(t,x) = E^x[g(X_\tau)].$$

Example. Let $d = 1$ and suppose that $B = (a,b)$ with $0 \le a < b < \infty$. Then $\partial B = \{a,b\}$. Take $a < x < b$ and consider

$$\tau = \inf\{t : X_t = a \text{ or } b\},$$

where X_t is a standard Brownian motion. Let g be the function on ∂B, $g(a) = 0, g(b) = 1$. Then

$$v(x) = E^x[g(X_\tau)] = P^x\{X_\tau = b\}$$

(here we have used P^x to denote a probability assuming $X_0 = x$). We know by above that $v(x)$ satisfies

$$\frac{d^2v}{dx^2} = 0, \quad a < x < b,$$

$$v(a) = 0, \quad v(b) = 1.$$

We can solve this differential equation easily and we get

$$v(x) = \frac{x-a}{b-a}.$$

This is the Brownian motion analogue of the gambler's ruin estimate.

8.5 Recurrence and Transience

In this section we ask whether the Brownian motion keeps returning to the origin. We have already answered this question for one-dimensional Brownian motion; if X_t is a standard (one-dimensional) Brownian motion, then X_t is *recurrent*, i.e., there are arbitrarily large times t with $X_t = 0$.

Now suppose X_t is a standard d-dimensional Brownian motion. Let $0 < R_1 < R_2 < \infty$ and let $B = B(R_1, R_2)$ be the annulus

$$B = \{x \in R^d : R_1 < |x| < R_2\},$$

with boundary

$$\partial B = \{x \in R^d : |x| = R_1 \text{ or } |x| = R_2\}.$$

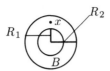

Suppose $x \in B$. Let $f(x) = f(x, R_1, R_2)$ be the probability that a standard Brownian motion starting at x hits the sphere $\{y : |y| = R_2\}$ before it hits the sphere $\{y : |y| = R_1\}$. If we let

$$\tau = \tau_{\partial B} = \inf\{t : X_t \in \partial B\},$$

then we can write

$$f(x) = E^x(g(X_\tau)),$$

where $g(y) = 1$ for $|y| = R_2$ and $g(y) = 0$ for $y = R_1$. We saw in the last section that f is the function satisfying

(i) $\Delta f(x) = 0$, $x \in B$,

(ii) $f(y) = 0$, $|y| = R_1$; $f(y) = 1$, $|y| = R_2$.

To find f, we first note that the symmetry of Brownian motion implies $f(x) = \phi(|x|)$ for some ϕ, i.e., the value of f depends only on the absolute value of x. We can write the equation (i) in spherical coordinates. The form of the Laplacian Δ in spherical coordinates is somewhat messy; however, it is not so bad for functions $\phi(r)$ that depend only on the radius. One can check that

$$\Delta\phi(r) = \frac{d^2\phi}{dr^2} + \frac{d-1}{r}\frac{d\phi}{dr}.$$

The general solution to the equation

$$\phi''(r) + \frac{d-1}{r}\phi'(r) = 0$$

is given by

$$\phi(r) = \begin{cases} c_1 \ln r + c_2, & d = 2, \\ c_1 r^{2-d} + c_2, & d \geq 3. \end{cases}$$

[The second-order equation for $\phi(r)$ is a first-order equation for $\psi(r) = \phi'(r)$ which can be solved by separation of variables.] Putting in the boundary conditions $\phi(R_1) = 0$ and $\phi(R_2) = 1$, we see that

$$f(x) = \phi(|x|) = \frac{\ln |x| - \ln R_1}{\ln R_2 - \ln R_1}, \quad d = 2,$$

$$f(x) = \phi(|x|) = \frac{R_1^{2-d} - |x|^{2-d}}{R_1^{2-d} - R_2^{2-d}}, \quad d \geq 3.$$

Consider now the two-dimensional case. Let $x \in R^2$ and suppose that a Brownian motion starts at x (or that the Brownian motion is at x at some time t). Take any $\epsilon > 0$, and ask the question: What is the probability that the Brownian motion never returns to the disc of radius ϵ about 0? The argument above gives us the probability of reaching the circle of radius R_2 before reaching the disc. The probability we are interested in is therefore

$$\lim_{R_2 \to \infty} \frac{\ln |x| - \ln \epsilon}{\ln R_2 - \ln \epsilon} = 0.$$

Hence, with probability 1 the Brownian motion always returns to the disc of radius ϵ and hence it returns infinitely often and at arbitrarily large times. Does it ever return to the point 0, i.e., are there times t with $X_t = 0$? Again, start the walk at $x \neq 0$. If there is a positive probability of reaching 0, then there must be an R_2 such that the probability of reaching 0 before reaching the circle of radius R_2 is positive. But this latter probability can be written as

$$\lim_{\epsilon \to 0} P^x\{|X_t| = \epsilon \text{ before } |X_t| = R_2\} = \lim_{\epsilon \to 0}[1 - \frac{\ln |x| - \ln \epsilon}{\ln R_2 - \ln \epsilon}] = 0.$$

Hence the Brownian motion never actually returns to 0. To summarize, the Brownian motion in two dimensions returns arbitrarily close to 0 infinitely often, but never actually returns to 0. We say that the Brownian motion in two dimensions is *neighborhood recurrent* but not *point recurrent*.

Now consider $d \geq 3$. Again we take $\epsilon > 0$ and ask what is the probability that the Brownian motion starting at x never returns to the ball of radius ϵ. If $|x| > \epsilon$, this is given by

$$\lim_{R_2 \to \infty} \frac{\epsilon^{2-d} - |x|^{2-d}}{\epsilon^{2-d} - R_2^{2-d}} = 1 - (\frac{\epsilon}{|x|})^{d-2} < 1.$$

Since the probability is less than 1, we can see that eventually the Brownian motion escapes from any ball around the origin and hence goes off to infinity. We say that in this case the Brownian motion is *transient*.

8.6 Fractal Nature of Brownian Motion

Let X_t be a standard d-dimensional Brownian motion and let A represent the (random) set of points visited by the path,

$$A = \{x \in R^d : X_t = x \text{ for some } t\}.$$

In this section we will consider the dimension of the set A for $d \geq 2$.

In order to consider a bounded set, let $A_1 = A \cap \{x : |x| \leq 1\}$. Fix an ϵ and let us try to cover A_1 with balls of diameter ϵ. First consider the whole ball of radius 1, $\{x : |x| \leq 1\}$ and cover it by balls of diameter ϵ. The number of such balls needed is of the order of ϵ^{-d} (which is consistent with the fact that the ball is a dimension d set). How many of these balls are needed to cover A_1?

First, consider $d = 2$. By the argument given in the previous section, every open ball is visited by the Brownian motion. Hence A intersects every ball and all the balls are needed. Hence the dimension of A is two.

Now consider $d > 2$. Take a typical ball of diameter ϵ. What is the probability that it is needed in the covering, i.e., what is the probability that Brownian motion visits the ball? By the calculations done in the previous section, a ball of radius $\epsilon/2$ around a point x (with $|x| > \epsilon/2$) is visited with probability $(\epsilon/2|x|)^{d-2}$. Hence, if ϵ is small and $|x|$ is of order 1, the probability is about a constant times ϵ^{d-2}. Since each of the about ϵ^{-d} balls is chosen for the covering with probability about ϵ^{d-2}, the total number of balls needed is about $\epsilon^{d-2}\epsilon^{-d} = \epsilon^{-2}$. Hence the dimension of the set A is two. We have just sketched the idea behind this following fact:

Fact. The path of a d-dimensional Brownian motion $(d \geq 2)$ has fractal dimension two.

8.7 Brownian Motion with Drift

Consider a d-dimensional Brownian motion X_t with variance parameter σ^2 starting at $x \in R^d$. Let $\mu \in R^d$ and

$$Y_t = X_t + t\mu.$$

Then Y_t is called d-dimensional Brownian motion with drift μ and variance parameter σ^2 starting at x. One can check easily that Y_t satisfies

(i) $Y_0 = x$;

(ii) if $s_1 \leq t_1 \leq s_2 \leq t_2 \leq \cdots \leq s_n \leq t_n$, then $Y_{t_1} - Y_{s_1}, \ldots, Y_{t_n} - Y_{s_n}$ are independent;

(iii) $Y_t - Y_s$ has a normal distribution with mean $\mu(t - s)$ and covariance matrix $\sigma^2(t - s)I$;

(iv) Y_t is a continuous function of t.

The motion Y_t consists of a "straight line" motion in the direction μ with random fluctuations. Note that $E(Y_t) = t\mu$.

The density of Y_t given $Y_0 = x$, $p_t(x, y)$ is easily seen to be

$$p_t(x, y) = \frac{1}{(2\pi\sigma^2 t)^{d/2}} e^{-|y-x-t\mu|^2/2t\sigma^2}.$$

This again satisfies the Chapman–Kolmogorov equation,

$$p_{s+t}(x, y) = \int_{R^d} p_s(x, z) p_t(z, y) \, dz_1 \cdots dz_d.$$

Suppose we start with a density on R^d, $f(x)$. Consider the function

$$f(t, x) = E^x[f(Y_t)].$$

For ease we will consider the case $d = 1, t = 0$. We again write f in a Taylor series about x,

$$f(y) = f(x) + f'(x)(y - x) + \frac{1}{2}f''(x)(y - x)^2 + o((y - x)^2).$$

Hence,

$$E^x[f(Y_t)] = f(x) + f'(x)E^x[Y_t - x]$$

$$+ \frac{1}{2}f''(x)E^x[(Y_t - x)^2] + o(E(Y_t - x)^2).$$

A Brownian motion with drift μ and variance parameter σ^2 starting at x can be obtained by letting $Y_t = X_t + t\mu + x$, where X_t is a (zero drift) Brownian motion with variance parameter σ^2 starting at 0. Hence,

$$E^x[Y_t - x] = E[X_t + t\mu] = t\mu,$$

$$E^x[(Y_t - x)^2] = E[(X_t + t\mu)^2] \quad = \quad [E(X_t + t\mu)]^2 + Var(X_t + t\mu)$$
$$= \quad (t\mu)^2 + \sigma^2 t.$$

Also, since $(Y_t - x)^2$ is order t, $o((Y_t - x)^2)$ is $o(t)$. Therefore,

$$\frac{\partial f}{\partial t}\Big|_{t=0} \quad = \quad \lim_{t \to \infty} \frac{E^x[f(Y_t)] - E^x[f(Y_0)]}{t}$$
$$= \quad \mu f'(x) + \frac{\sigma^2}{2}f''(x).$$

We see that the inclusion of a drift has added a first derivative with respect to x.

In d dimensions, if the drift $\mu = (\mu_1, \ldots, \mu_d)$, we would get

$$\frac{\partial f}{\partial t} = \sum_{i=1}^{d} \mu_i \frac{\partial f}{\partial x_i} + \frac{\sigma^2}{2}\Delta f.$$

8.8 Exercises

8.1 Let X be a normal random variable, mean 0 variance 1. Show that if $a > 0$

$$P\{X \geq a\} \leq \frac{2}{a\sqrt{2\pi}}e^{-a^2/2}.$$

(Hint:

$$\int_a^\infty e^{-x^2/2}dx \leq \int_a^\infty e^{-ax/2}\, dx. \quad)$$

8.2 Let $X_{n1}, \ldots, X_{nn}$ be independent normal random variables with mean 0 and variance $1/n$. Then

$$X = X_{n1} + \cdots + X_{nn},$$

is a normal random variable with mean 0, variance 1. Let

$$M_n = \max\{|X_{n1}|, \ldots, |X_{nn}|\}.$$

Show that for every $\epsilon > 0$,

$$\lim_{n \to \infty} P\{M_n > \epsilon\} = 0.$$

(Hint: it will be useful to use the estimate from Problem 8.1. It may also be useful to remember that if Y is normal mean 0, variance σ^2, then $\sigma^{-1}Y$ is normal mean 0, variance 1.)

8.3 Let $X_{n1}, \ldots, X_{nn}$ be independent Poisson random variables with mean $1/n$. Then

$$X = X_{n1} + \cdots + X_{nn},$$

is a Poisson random variable with mean 1. Let

$$M_n = \max\{X_{n1}, \ldots, X_{nn}\}.$$

What is

$$\lim_{n \to \infty} P\{M_n > 1/2\} \quad ?$$

8.4 Random variables $Y_1, \ldots, Y_n$ have a joint normal distribution with mean 0 if there exist independent normal random variables $X_1, \ldots, X_n$, each normal mean 0, variance 1, and constants a_{ij} such that

$$Y_i = a_{i1}X_1 + \cdots + a_{in}X_n.$$

Let X_t be a standard Brownian motion. Let $s_1 \leq s_2 \leq \cdots \leq s_n$. Explain why it follows from the definition of a Brownian motion that $X_{s_1}, \ldots, X_{s_n}$ have a joint normal distribution.

8.5 The covariance matrix for joint normal random variables $Y_1, \ldots, Y_n$ is the matrix Γ whose (i, j) entry is $E(Y_i Y_j)$. Let X_t and $s_1, \ldots, s_n$ be as in Exercise 8.4. Find the covariance matrix Γ for $X_{s_1}, \ldots, X_{s_n}$.

8.6 Suppose X_t is a standard Brownian motion and $Y_t = a^{-1/2}X_{at}$ with $a > 0$. Show that Y_t is a standard Brownian motion.

8.7 Suppose X_t is a standard Brownian motion and $Y_t = tX_{1/t}$. Show that Y_t is a standard Brownian motion.

8.8 Let X_t be a standard Brownian motion. Compute the following conditional probability:

$$P\{X_2 > 0 \mid X_1 > 0\}.$$

Are the events $\{X_1 > 0\}$ and $\{X_2 > 0\}$ independent?

8.9 Let X_t and Y_t be independent standard (one dimensional) Brownian motions.

(a) Show that $Z_t = X_t - Y_t$ is a Brownian motion. What is the variance parameter for Z_t?

(b) True or False: With probability 1, $X_t = Y_t$ for infinitely many values of t.

8.10 Let the Cantor-like set A be defined as follows. Let $A_0 = [0, 1]$,

$$A_1 = [0, \frac{2}{5}] \cup [\frac{4}{5}, 1],$$

and A_n is obtained from A_{n-1} by removing the "middle fifth" from each interval in A_{n-1}. Let

$$A = \cap_{n=0}^{\infty} A_n.$$

What is the fractal dimension of A?

CHAPTER 9

Stochastic Integration

9.1 Integration with Respect to Random Walk

The goal of this chapter is to introduce the idea of integration with respect to Brownian motion. To give the reader a sense for the integral, we will start by discussing integration with respect to simple random walk. Let $X_1, X_2, \ldots$ be independent random variables, $P\{X_i = 1\} = P\{X_i = -1\} = 1/2$ and let S_n denote the corresponding simple random walk

$$S_n = X_1 + \cdots + X_n.$$

As in Chapter 5, Section 5.2, Example 3, we think of X_n as being the result of a game at time n and we can consider possible betting strategies on the games.

Let $\mathcal{F}_n$ denote the information contained in $X_1, \ldots, X_n$. Let B_n be the "bet" on the nth game. B_n can be either positive or negative, a negative value being the same as betting that X_n will turn up -1. The important assumption that we make is that the bettor must make the bet using only the information available up to, but not including, the nth game, i.e., we assume that B_n is measurable with respect to $\mathcal{F}_{n-1}$. The winnings up to time n, Z_n, can be written as

$$Z_n = \sum_{i=1}^{n} B_i X_i = \sum_{i=1}^{n} B_i [S_i - S_{i-1}] = \sum_{i=1}^{n} B_i \Delta S_i,$$

where we write $\Delta S_i = S_i - S_{i-1}$. We call Z_n the integral of B_n with respect to S_n.

There are two important properties that this integral satisfies. The first was shown in Chapter 5, Section 5.2, Example 3: the process Z_n is a martingale with respect to $\mathcal{F}_n$, i.e., if $m < n$,

$$E(Z_n \mid \mathcal{F}_m) = Z_m.$$

In particular, $E(Z_n) = 0$. The second property deals with the second moment of Z_n. Assume that the bets B_n have finite second moments, i.e.,

$E(B_n^2) < \infty$. Then

$$Var(Z_n) = E(Z_n^2) = \sum_{i=1}^{n} E(B_i^2).$$

To see this, we write

$$Z_n^2 = \sum_{i=1}^{n} B_i^2 X_i^2 + 2 \sum_{1 \le i < j \le n} B_i B_j X_i X_j.$$

Note that $X_i^2 = 1$ and hence

$$E(\sum_{i=1}^{n} B_i^2 X_i^2) = \sum_{i=1}^{n} E(B_i^2).$$

Suppose $i < j$. Then B_i, X_i, B_j are all measurable with respect to $\mathcal{F}_{j-1}$ while X_j is independent of $\mathcal{F}_{j-1}$. Using (5.3), we see that

$$E(B_i B_j X_i X_j \mid \mathcal{F}_{j-1}) = B_i B_j X_i E(X_j \mid \mathcal{F}_{j-1}) = B_i B_j X_i E(X_j) = 0,$$

and hence

$$E(B_i B_j X_i X_j) = E[E(B_i B_j X_i X_j \mid \mathcal{F}_{j-1})] = 0.$$

9.2 Integration with Respect to Brownian Motion

Here we describe a continuous analogue of the discrete integral given in the last section. Instead of a simple random walk, we will take a standard (one-dimensional) Brownian motion, which we will write W_t. We can think of this as a continuous fair game such that if one bets one unit for the entire period $[s, t]$ then one's winnings in this time period would be $W_t - W_s$.

Let Y_t denote the amount that is bet at time t. What we would like to do is define

$$Z_t = \int_0^t Y_s \, dW_s.$$

The process Z_t should denote the amount won in this game up to time t if the amount bet at time s is Y_s. It is a nontrivial mathematical problem to define this integral. The roughness of the paths of the Brownian motion prevent one from defining the integral as a "Riemann–Stieljes" integral.

We will make two assumptions about our betting strategy Y_s. The first assumption is that $E(Y_t^2) < \infty$ for all t and for each t,

$$\int_0^t E(Y_s^2) \, ds < \infty.$$

This condition will certainly be satisfied if we restrict ourselves to bounded betting strategies. The second assumption is critical and corresponds to our assumption before that the bettor cannot look into the future to determine the bet. Let $\mathcal{F}_t$ denote the information contained in the Brownian motion up through time t. We assume that Y_t is $\mathcal{F}_t$ measurable. In other words, the bettor can see the entire Brownian motion up through time t before choosing the bet, but cannot see anything after time t.

It is not too difficult to define the integral if we make the restrictive assumption that the bettor can change the bet only at a certain finite set of times, say $t_1 < t_2 < \cdots < t_n$. The bets then take the form

$$Y_t = \begin{cases} Y_0 & 0 \le t < t_1 \\ Y_1 & t_1 \le t < t_2 \\ \vdots \\ Y_n & t_n \le t < \infty. \end{cases}$$

Here $Y_0, \ldots, Y_n$ are random variables with $E(Y_i^2) < \infty$, and Y_i must be measurable with respect to $\mathcal{F}_{t_i}$ (where $t_0 = 0$). We will call a betting strategy that can change at only a finite number of times a simple strategy. For a simple strategy, we define the stochastic integral for $t_j \le t < t_{j+1}$ by

$$Z_t = \int_0^t Y_s \, dW_s = \sum_{i=1}^j Y_{i-1}[W_{t_i} - W_{t_{i-1}}] + Y_j[W_t - W_{t_j}].$$

There are three important properties that the stochastic integral of a simple strategy satisfies. The first is linearity: if X_s and Y_s are two simple strategies and a, b are real numbers, then $aX_s + bY_s$ is a simple strategy and

$$\int_0^t (aX_s + bY_s) \, dW_s = a \int_0^t X_s \, dW_s + b \int_0^t Y_s \, dW_s.$$

This can be easily checked.

The other two properties are direct analogues of the properties of the discrete stochastic integral of the previous section. We say a continuous-time process Z_t is a martingale with respect to $\mathcal{F}_t$ if each Z_t is $\mathcal{F}_t$ measurable; $E(|Z_t|) < \infty$ for each t; and if $s < t$,

$$E(Z_t \mid \mathcal{F}_s) = Z_s. \tag{9.1}$$

The second property is that the stochastic integral Z_t as defined above is a martingale with respect to the information $\mathcal{F}_t$ derived from the Brownian motion. It is easy to see that Z_t is $\mathcal{F}_t$ measurable and the condition $E(|Z_t|) < \infty$ follows from the fact that the second moments of the Y_i exist. We will now verify (9.1). First assume $t_j \le s < t \le t_{j+1}$ for some j. Then we can write

$$Z_t = Z_s + Y_j[W_t - W_s].$$

Since Y_j and Z_s are $\mathcal{F}_s$ measurable and $W_t - W_s$ is independent of $\mathcal{F}_s$,

$$E(Z_t \mid \mathcal{F}_s) = Z_s + Y_j E(W_t - W_s \mid \mathcal{F}_s) = Z_s + Y_j E(W_t - W_s) = Z_s.$$

In particular, if $t_j \le t \le t_{j+1}$,

$$E(Z_{t_{j+1}} \mid \mathcal{F}_t) = Z_t, \quad E(Z_t \mid \mathcal{F}_{t_j}) = Z_{t_j}.$$

Note that $E(Z_t \mid \mathcal{F}_{t_{j-1}}) = E(E(Z_t \mid \mathcal{F}_{t_j}) \mid \mathcal{F}_{t_{j-1}}) = E(Z_{t_j} \mid \mathcal{F}_{t_{j-1}}) = Z_{t_{j-1}}$, and by iteration we can see that for all $i \le j$, $E(Z_t \mid \mathcal{F}_{t_i}) = Z_{t_i}$. Finally, if $t_i \le s < t_{i+1}, t_j \le t < t_{j+1}$ for some $i < j$, then

$$E(Z_t \mid \mathcal{F}_s) = E(E(Z_t \mid \mathcal{F}_{t_{i+1}}) \mid \mathcal{F}_s) = E(Z_{t_{i+1}} \mid \mathcal{F}_s) = Z_s.$$

This gives (9.1).

The third property gives a way to calculate the second moment,

$$E(Z_t^2) = \int_0^t E(Y_s^2) \, ds. \tag{9.2}$$

The right-hand side is a standard calculus "ds" integral. To prove this, assume that $t_j \le t < t_{j+1}$. Note that $E(Y_s^2)$ is a step function in s so

$$\int_0^t E(Y_s^2) \, ds = \sum_{i=0}^{j-1} E(Y_i^2)(t_{i+1} - t_i) + E(Y_j^2)(t - t_j).$$

If we expand the square, we see that

$$Z_t^2 = \sum_{i=1}^{j} Y_{i-1}^2 [W_{t_i} - W_{t_{i-1}}]^2 + Y_j^2 [W_t - W_{t_j}]^2 + \text{(cross terms)},$$

where "cross terms" represents a sum of terms of the form

$$Y_{i-1} Y_{k-1} [W_{t_i} - W_{t_{i-1}}][W_{t_k} - W_{t_{k-1}}], \quad i < k,$$

or

$$Y_{i-1} Y_j [W_{t_i} - W_{t_{i-1}}][W_t - W_j].$$

If $i < k$,

$$
\begin{aligned}
E(Y_{i-1} Y_{k-1} [W_{t_i} & - W_{t_{i-1}}][W_{t_k} - W_{t_{k-1}}] \mid \mathcal{F}_{t_{k-1}}) \\
&= Y_{i-1} Y_{k-1} [W_{t_i} - W_{t_{i-1}}] E(W_{t_k} - W_{t_{k-1}} \mid \mathcal{F}_{t_{k-1}}) \\
&= Y_{i-1} Y_{k-1} [W_{t_i} - W_{t_{i-1}}] E(W_{t_k} - W_{t_{k-1}}) \\
&= 0,
\end{aligned}
$$

and hence

$$E(Y_{i-1} Y_{k-1} [W_{t_i} - W_{t_{i-1}}][W_{t_k} - W_{t_{k-1}}]) =$$
$$E[E(Y_{i-1} Y_{k-1} [W_{t_i} - W_{t_{i-1}}][W_{t_k} - W_{t_{k-1}}] \mid \mathcal{F}_{t_{k-1}})] = 0.$$

Similarly,

$$E(Y_{i-1} Y_j [W_{t_i} - W_{t_{i-1}}][W_t - W_j]) = 0.$$

Therefore

$$E(Z_t^2) = \sum_{i=1}^{j} E(Y_{i-1}^2 [W_{t_i} - W_{t_{i-1}}]^2) + E(Y_j^2 [W_t - W_{t_j}]^2).$$

Note that

$$
\begin{aligned}
E[Y_{i-1}^2[W_{t_i} - W_{t_{i-1}}]^2 \mid \mathcal{F}_{t_{i-1}}] &= Y_{i-1}^2 E[(W_{t_i} - W_{t_{i-1}})^2 \mid \mathcal{F}_{t_{i-1}}] \\
&= Y_{i-1}^2 E[(W_{t_i} - W_{t_{i-1}})^2] \\
&= Y_{i-1}^2 (t_i - t_{i-1}).
\end{aligned}
$$

Hence,

$$
\begin{aligned}
E[Y_{i-1}^2[W_{t_i} - W_{t_{i-1}}]^2] &= E(E[Y_{i-1}^2[W_{t_i} - W_{t_{i-1}}]^2 \mid \mathcal{F}_{t_{i-1}}]) \\
&= E(Y_{i-1}^2)(t_i - t_{i-1}).
\end{aligned}
$$

Similarly,

$$
E[Y_j^2[W_t - W_{t_j}]^2] = E(Y_j^2)(t - t_j).
$$

This proves (9.2).

To define the stochastic integral for betting rules Y_s that are not simple, we do the standard mathematical procedure for defining continuous objects—approximate by discrete and take a limit. Let Y_s be measurable with respect to $\mathcal{F}_s$, satisfying the second moment conditions listed above. A little more must be assumed about the Y_s to be mathematically precise: the paths of Y_s (i.e., Y_s considered as a function of s) should be right continuous and have left limits; we will not worry about this in our informal treatment. For each $n \geq 0$, define the approximate stategy $Y_s^{(n)}$ by

$$
Y_s^{(n)} = n \int_{(k-1)/n}^{k/n} Y_r \, dr, \quad \frac{k}{n} < s \leq \frac{k+1}{n},
$$

where we set $Y_s^{(n)} = 0$ for $s \leq 1/n$. We have arranged the approximation so that for each t, $Y_s^{(n)}$, $0 \leq s \leq t$ is a simple strategy that is $\mathcal{F}_s$ measurable. The key estimate that can be proved (we will not do it) is that

$$
Y_s^{(n)} \to Y_s
$$

in the sense that for each t

$$
\lim_{n \to \infty} \int_0^t E([Y_s - Y_s^{(n)}]^2) \, ds = 0.
$$

This allows us to define the stochastic integral

$$
Z_t = \int_0^t Y_s \, dW_s,
$$

by saying that Z_t is the mean square limit of the random variables

$$
Z_t^{(n)} = \int_0^t Y_s^{(n)} \, dW_s.
$$

The first and third properties of the stochastic integral allow this definition to work since as $n, m \to \infty$,

$$E([Z_t^{(n)} - Z_t^{(m)}]^2) = \int_0^t E([Y_s^{(n)} - Y_s^{(m)}]^2) \, ds \to 0.$$

In the process of showing the limit exists, one also shows that the three properties of the integral still hold.

Linearity:

$$\int_0^t [aX_s + bY_s] \, dW_s = a \int_0^t X_s \, dW_s + b \int_0^t Y_s \, dW_s.$$

Martingale Property: $Z_t = \int_0^t Y_s \, dW_s$ is a martingale with respect to $\mathcal{F}_t$. In particular, $E(Z_t) = 0$ for all t.

Second Moment Calculation:

$$Var\left(\int_0^t Y_s \, dW_s\right) = E\left[\left(\int_0^t Y_s \, dW_s\right)^2\right] = \int_0^t E[Y_s^2] \, ds.$$

The relationship

$$Z_t = \int_0^t Y_s \, dW_s$$

is often written in the differential form

$$dZ_t = Y_t \, dW_t.$$

The process Z_t can be thought of as a process that at time t looks like a Brownian motion with variance parameter Y_t^2 (recall that if W_t is a standard Brownian motion, then σW_t is a Brownian motion with variance parameter σ^2.) Sometimes one has a process

$$Z_t = \int_0^t X_s \, ds + \int_0^t Y_s \, dW_s,$$

where the "ds" integral is a standard calculus integral. In differential form this is written

$$dZ_t = X_t \, dt + Y_t \, dW_t.$$

This represents a process that at time t looks like a Brownian motion with drift X_t and variance parameter Y_t^2.

9.3 Ito's Formula

How does one calculate stochastic integrals? As an example, consider the integral

$$Z_t = \int_0^t W_s \, dW_s.$$

W_s is $\mathcal{F}_s$ measurable and this integral is well defined. One might hope that standard calculus rules would work for stochastic integrals in which case we would have

$$\int_0^t W_s \, dW_s = \frac{1}{2}W_t^2 - \frac{1}{2}W_0^2 = \frac{1}{2}W_t^2.$$

However, a quick examination of this equation shows that it cannot be true: the left side is a random variable with expectation 0 but the right-hand side has expectation $t/2$. In this section, we derive a formula that will allow us to calculate this integral exactly. This formula is usually called Ito's formula and it is the fundamental theorem of stochastic calculus.

Let us start by reviewing the ordinary fundmental theorem of calculus. Suppose we have a continuously differential function $f(t)$. Around each t_0 we can expand $f(t)$,

$$f(t) = f(t_0) + f'(t_0)(t - t_0) + o(t - t_0).$$

We can write $f(t)$ as a telescoping sum

$$f(t) = f(0) + \sum_{j=0}^{n-1}[f(\frac{(j+1)t}{n}) - f(\frac{jt}{n})].$$

We now use the Taylor's series about jt/n to write

$$f(\frac{(j+1)t}{n}) = f(\frac{jt}{n}) + f'(\frac{jt}{n})\frac{t}{n} + to(\frac{1}{n}),$$

and

$$f(t) - f(0) = \sum_{j=0}^{n-1} f'(\frac{jt}{n})\frac{t}{n} + \sum_{j=0}^{n-1} to(\frac{1}{n}).$$

As $n \to \infty$ the second term on the right tends to 0 and the first term tends to the integral of f'. We therefore get

$$f(t) - f(0) = \int_0^t f'(s) \, ds,$$

which we all know very well.

Now let W_t be a Brownian motion, and f a function with at least two continuous derivatives. At each x_0 we can expand $f(x)$,

$$f(x) = f(x_0) + f'(x_0)(x - x_0) + \frac{1}{2}f''(x_0)(x - x_0)^2 + o((x - x_0)^2).$$

Write $f(W_t)$ as a telescoping sum,

$$f(W_t) = f(W_0) + \sum_{j=0}^{n-1}[f(W_{\frac{i+1}{n}t}) - f(W_{\frac{i}{n}t})].$$

By using the Taylor series expansion about $W_{\frac{j}{n}t}$ we can write

$$f(W_{\frac{j+1}{n}t}) = f(W_{\frac{j}{n}t}) + f'(W_{\frac{j}{n}t})[W_{\frac{j+1}{n}t} - W_{\frac{j}{n}t}]$$

$$+ \frac{1}{2}f''(W_{\frac{j}{n}t})[W_{\frac{j+1}{n}t} - W_{\frac{j}{n}t}]^2 + to(\frac{1}{n}).$$

The $o(\cdot)$ is smaller than order n^{-1} since $[W_{\frac{j+1}{n}t} - W_{\frac{j}{n}t}]^2$ is of order n^{-1}. We then get

$$f(W_t) - f(W_0) = \sum_{j=0}^{n-1} f'(W_{\frac{j}{n}t})[W_{\frac{j+1}{n}t} - W_{\frac{j}{n}t}]$$

$$+ \frac{1}{2}\sum_{j=0}^{n-1} f''(W_{\frac{j}{n}t})[W_{\frac{j+1}{n}t} - W_{\frac{j}{n}t}]^2 + \sum_{j=0}^{n-1} to(\frac{1}{n}). \qquad (9.3)$$

As $n \to \infty$, the third term on the right goes to 0. Since f' is continuous, the first term will approach

$$\int_0^t f'(W_s)\, dW_s.$$

To see what the second term converges to, let us consider the general question of the limit of

$$\sum_{j=0}^{n-1} g(W_{\frac{j}{n}t})[W_{\frac{j+1}{n}t} - W_{\frac{j}{n}t}]^2,$$

where g is a continuous function. First consider the case where g is identically 1. Let

$$Q_t^{(n)} = \sum_{j=0}^{n-1} [W_{\frac{j+1}{n}t} - W_{\frac{j}{n}t}]^2.$$

The limit

$$Q_t = \lim_{n\to\infty} Q_t^{(n)}$$

is often called the quadratic variation of W_t. $[W_{\frac{j+1}{n}t} - W_{\frac{j}{n}t}]^2$ has the same distribution as $(t/n)U^2$, where U is normal mean 0, variance 1. Note that

$$E(U^2) = 1 \quad Var(U^2) = E(U^4) - [E(U^2)]^2 = 2.$$

Hence, since the increments of W are independent,

$$E(Q_t^{(n)}) = \sum_{j=0}^{n-1} E([W_{\frac{j+1}{n}t} - W_{\frac{j}{n}t}]^2) = t,$$

$$Var(Q_t^{(n)}) = \sum_{j=0}^{n-1} Var([W_{\frac{j+1}{n}t} - W_{\frac{j}{n}t}]^2) = nVar((t/n)U^2) = \frac{2t^2}{n}.$$

As $n \to \infty$, the expectation of $Q_t^{(n)}$ stays constant but the variance goes to 0. In other words, the limiting random variable Q_t is just a constant, and the quadratic variation of Brownian motion up to time t is the constant random variable equal to t.

For any g let

$$Q_t^{(n)}(g) = \sum_{j=0}^{n-1} g(W_{\frac{j}{n}t})[W_{\frac{j+1}{n}t} - W_{\frac{j}{n}t}]^2,$$

and

$$Q_t(g) = \lim_{n \to \infty} Q_t^{(n)}(g).$$

If g is a step function of the form

$$g(s) = u(W_{\frac{j}{m}t}), \quad \frac{j}{m}t \le s < \frac{j+1}{m}t,$$

then

$$
\begin{aligned}
Q_t(g) &= \lim_{n \to \infty} Q_t^{(n)}(g) \\
&= \lim_{k \to \infty} Q_t^{(km)}(g) \\
&= \lim_{k \to \infty} \sum_{j=0}^{m-1} u(W_{\frac{j}{m}t}) \sum_{i=0}^{k-1} [W_{\frac{kj+i+1}{km}t} - W_{\frac{kj+i}{km}t}]^2 \\
&= \sum_{j=0}^{m-1} u(W_{\frac{j}{m}t}) \lim_{k \to \infty} \sum_{i=0}^{k-1} [W_{\frac{kj+i+1}{km}t} - W_{\frac{kj+i}{km}t}]^2.
\end{aligned}
$$

The result about quadratic variation tells us that

$$\lim_{k \to \infty} \sum_{i=0}^{k-1} [W_{\frac{kj+i+1}{km}t} - W_{\frac{kj+i}{km}t}]^2 = \frac{t}{m}.$$

Hence

$$Q_t(g) = \sum_{j=0}^{m-1} u(W_{\frac{j}{m}t}) \frac{t}{m}.$$

Now assume g is continuous. For each n, let g_n be the step function

$$g_n(s) = g(W_{\frac{j}{n}t}), \quad \frac{j}{n}t \le s < \frac{j+1}{n}t.$$

Note that

$$|Q_t(g) - Q_t(g_n)| \le \|g - g_n\| Q_t = t\|g - g_n\|,$$

where

$$\|g - g_n\| = \sup_{0 \le s \le t} |g(W_s) - g_n(s)|.$$

The continuity of g and the continuity of the Brownian motion imply that $\|g - g_n\| \to 0$ as $n \to \infty$. Hence

$$
\begin{aligned}
Q_t(g) &= \lim_{n\to\infty} Q_t(g_n) \\
&= \lim_{n\to\infty} \sum_{j=0}^{n-1} g(\tfrac{j}{n} t) \tfrac{t}{n}.
\end{aligned}
$$

The last expression is the usual representation of the integral of g as a limit of Riemann sums. Therefore, if g is continuous,

$$
Q_t(g) = \int_0^t g(s)\, ds.
$$

If we plug this result into (9.3) we can conclude the following.

Ito's Formula. If f is a function with two continuous derivatives, and W_t is a standard Brownian motion,

$$
f(W_t) - f(W_0) = \int_0^t f'(W_s)\, dW_s + \frac{1}{2}\int_0^t f''(W_s)\, ds.
$$

This formula is sometimes written in the differential form,

$$
df(W_t) = f'(W_t)\, dW_t + \frac{1}{2}f''(W_t)\, dt.
$$

As an example, let $f(t) = t^2$. Then $f'(t) = 2t$, $f''(t) = 2$, and

$$
W_t^2 = \int_0^t 2W_s\, dW_s + \frac{1}{2}\int_0^t 2W_s\, ds,
$$

or

$$
\int_0^t W_s\, dW_s = \frac{1}{2}W_t^2 - \frac{1}{2}t.
$$

This turns out to be a particularly nice example; in general, one cannot use Ito's formula to calculate integrals exactly.

As another example, consider the process

$$
X_t = e^{W_t}.
$$

This process is often called geometric Brownian motion. It is a natural model for modelling stock prices. Ito's formula with $f(t) = e^t$ says that

$$
X_t - 1 = \int_0^t e^{W_s}\, dW_s + \frac{1}{2}\int_0^t e^{W_s}\, ds.
$$

In other words X_t satisfies the stochastic differential equation

$$
dX_t = X_t\, dW_t + \frac{1}{2}X_t\, dt.
$$

9.4 Simulation

Consider a stochastic differential equation

$$dX_t = a(X_t)\, dt + b(X_t)\, dW_t,$$

where a and b are relatively nice functions of x and W_t denotes a standard Brownian motion. The solution is a process X_t that at any particular time looks like a Brownian motion with drift parameter $a(X_t)$ and variance parameter $b(X_t)$. While it is often difficult to give an explicit solution to the equation, it is easy to simulate the process on a computer using a random walk.

Choose some small number Δt. We can approximate the Brownian motion by a simple random walk with time increments Δt and space increments $\sqrt{\Delta t}$. To do this let $Y_1, Y_2, \ldots$ be independent random variables with

$$P\{Y_i = 1\} = P\{Y_i = -1\} = \frac{1}{2}.$$

We set $X_0 = 0$ and for $n > 0$,

$$X_{n\Delta t} = X_{(n-1)\Delta t} + a(X_{(n-1)\Delta t})\Delta t + b(X_{(n-1)\Delta t})\sqrt{\Delta t}\, Y_n.$$

9.5 Exercises

9.1 Let X_t be a standard (one-dimensional) Brownian motion. As before let $\mathcal{F}_t$ denote the information contained in $\{X_s : 0 \le s \le t\}$, i.e., all the information contained in the path up through time t. We call a continuous-time process M_t a martingale with respect to $\mathcal{F}_t$ if for all $s < t$,

$$E(M_t \mid \mathcal{F}_s) = M_s.$$

Which of the following are martingales with respect to $\mathcal{F}_t$?
  (a) $M_t = X_t$;
  (b) $M_t = X_t^2$;
  (c) $M_t = X_t^2 - t$;
  (d) $M_t = e^{X_t - (t^2/2)}$
  (e) $M_t = X_{\tau_t}$, where $\tau = \inf\{t : X_t = 1\}$ and $\tau_t = \min\{t, \tau\}$.

9.2 In the same way that Ito's formula is derived, a slightly more general version can be proved. Suppose $f(t, x)$ is a function of two variables that has two continuous derivatives in x and at least one continuous derivative in t. Then

$$f(t, W_t) - f(0, W_0) = \int_0^t f_s(s, W_s)\, ds$$

$$+ \int_0^t f_x(s, W_s)\, dW_s + \frac{1}{2}\int_0^t f_{xx}(s, W_s)\, ds,$$

where f_s denotes differentiation in the first variable and f_x, f_{xx} denote differentiation in the second variable. In differential form this becomes

$$df(t, W_t) = [f_t(t, W_t) + \frac{1}{2}f_{xx}(t, W_t)] \, dt + f_x(t, W_t) \, dW_t.$$

Let W_t be a standard Brownian motion and let

$$X_t = e^{at+bW_t}.$$

Show that X_t satisfies the stochastic differential equation

$$dX_t = (a + \frac{1}{2}b^2)X_t \, dt + bX_t \, dW_t.$$

For which values of a and b is X_t a martingale?

9.3 COMPUTER SIMULATION. Assume X_t is a process satisfying the stochastic differential equation

$$dX_t = a(X_t) \, dt + b(X_t) \, dW_t,$$

where

$$a(x) = 0,$$

$$b(x) = \begin{cases} 2, & x > 0 \\ 1 & x < 0. \end{cases}$$

Using $\Delta t = 1/100$ run many simulations of X_t. Estimate the following
   (a) $E(X_1)$,
   (b) $P\{X_1 > 0\}$

9.4 Do Exercise 9.3 with

$$a(x) = x,$$

$$b(x) = \sqrt{|x|}.$$

Index